Proceedings of the Second International Firefly Symposium, Selangor, Malaysia 2010

Firefly Conservation: From Science to Practice

Proceedings of the Second International Firefly Symposium

Selangor, Malaysia 2010

LAURENCE G. KIRTON, *Editor*
GRACE T. LIM, *Editor*
Forest Research Institute Malaysia, Kepong
JOHN C. DAY, *Editor*
Centre for Ecology & Hydrology (CEH), Wallingford

LAMPYRID | Volume 2 | 2012

Brazen Head Publishing
Lombardy House, Thornham, Norfolk, PE36 6LX, United Kingdom

www.brazenheadpublishing.co.uk

Published in the United Kingdom
by Brazen Head Publishing

First published 2012

Paperback edition published 2014

International Standard Serial Number: 2041-4900 (print)
International Standard Book Number: 978-0-9565106-4-8 (Hardcover)
International Standard Book Number: 978-0-9565106-7-9 (Paperback)

Contents

Poster Abstracts

Foreword

THE SECOND INTERNATIONAL FIREFLY SYMPOSIUM, held in Malaysia from 2–5 August 2010, was a landmark event held in conjunction with the International Year for Biodiversity 2010. Organised primarily by the Forest Research Institute Malaysia (FRIM), it brought together seventy four participants from thirteen countries, including scientists, members of the public, the media, and representatives of government and non-government agencies who shared a common interest in fireflies, their conservation and sustainable ecotourism. It is therefore with great pleasure that I present the proceedings of this symposium, which publishes the findings presented at the symposium in a special volume of the international journal of bioluminescent beetle research, Lampyrid.

Fireflies stand out among the millions of insect species that inhabit our Earth because of their fascinating ability to produce light. The physiology of their bioluminescence has been much studied, and has even been tapped as a tool in the study of cell processes in the medical and life sciences. Fireflies have also become indicators of the quality of our rivers and countryside. Ecotourism industries are starting to flourish around firefly populations, as clearly seen in Malaysia's famed Firefly Park along the Selangor River estuary. Fireflies have also been a source of inspiration in the arts. This, as well as their iconic nature, makes them good flagship species for conservation awareness and advocacy.

The Forest Research Institute Malaysia has for many years led research in the conservation of the Selangor River fireflies, determining their habitat requirements and developing a monitoring technique and programme for the population, as well as working with the government to protect its habitat. These proceedings represent yet another contribution of the Institute to our knowledge of fireflies and their conservation biology. The proceedings are unique in that contributions come not only from scientists but also from firefly enthusiasts of various backgrounds. To all contributors, and the editors who endeavoured so greatly to enable each contribution to be published as a scientific paper, I extend my appreciation. We are very grateful to John C. Day for the opportunity to publish the proceedings in the journal Lampyrid.

To all the organisations and agencies that worked with us or supported the symposium and these proceedings financially—the Malaysian Forestry Research and Development Board, Selangor State Government, Tourism Selangor, Ministry of Tourism Malaysia, Tenaga Nasional Berhad, Malaysian Nature Society, Malakoff Corporation, Selangor Waters Management Authority, and Kuala Selangor District Council—I also extend my heartfelt thanks.

Dato' Dr. Abd. Latif Mohmod
Director General, Forest Research Institute Malaysia (FRIM)

Preface

THIS SPECIAL VOLUME OF THE JOURNAL LAMPYRID is devoted to the Proceedings of the Second International Firefly Symposium held in Holiday Villa Hotel and Suites in Subang, Selangor, Malaysia from 2–5 August 2010. The symposium was preceded by a training course on Identification and Taxonomy of South East Asian Fireflies held in the Forest Research Institute Malaysia from 27–31 July 2010 and conducted by two respected firefly taxonomists, Lesley Ballantyne (Charles Sturt University, Australia) and Jeng Ming-Luen (National Museum of Natural Sciences, Taiwan).

The theme of the symposium was "Firefly Conservation: From Science to Practice." In keeping with the theme, a declaration named after the state of Selangor was made by participants at the Third International Firefly Research Network Meeting held on the final day of the symposium. The state of Selangor is where the heart of Malaysian firefly ecotourism lies. The Selangor Declaration on Conservation of Fireflies, which is published in these proceedings, advocates the importance of understanding firefly taxonomy, diversity and behaviour as a basis for conservation efforts. It also stresses the role that governments, local authorities, agencies and corporations can play in helping to conserve fireflies and in raising awareness.

A total of twenty six oral papers and ten posters were presented at the symposium under the themes of Taxonomy and Phylogeny, Biology and Behaviour, Ecology and Diversity, and Conservation and Ecotourism. Of these, twenty papers are published in these proceedings. Authors of the remaining papers have chosen to publish the work they presented in other journals. However, abstracts of their presentations are published in these proceedings, as well as abstracts of posters presented. All papers were reviewed by the editors, but we are grateful to have had the assistance of Lesley A. Ballantyne (Charles Sturt University) and Joseph M. Cicero (University of Arizona), who reviewed specific papers in their areas of expertise, and Chooi-Khim Phon, who provided valuable assistance in proof reading.

The symposium was made possible through the organisational efforts of the Forest Research Institute Malaysia and Malaysian Nature Society. Financial support for the organisation of the symposium and publishing of these proceedings was provided by the Malaysian Forestry Research and Development Board, Selangor State Government, Tourism Selangor, Ministry of Tourism Malaysia, Tenaga Nasional Berhad, Malaysian Nature Society, Malakoff Corporation, Selangor Waters Management Authority, and Kuala Selangor District Council.

Laurence G. Kirton
Forest Research Institute Malaysia

The Selangor Declaration
Conservation of Fireflies

Participants from 13 countries met at the Second International Firefly Symposium in Selangor, Malaysia from 2nd to 5th August 2010. They included experts in the fields of taxonomy, genetics, biology, behaviour, ecology and conservation of fireflies as well as members of government agencies, non-governmental organisations, educational institutions, corporations and the public. The theme of the symposium was 'Firefly Conservation: From Science to Practice.' The following declaration was made in conjunction with the symposium at the Third International Firefly Research Network Meeting on 5th August 2010.

THE ABILITY OF FIREFLIES TO PRODUCE LIGHT has been used to the benefit mankind in biomedical research, yet little is known of the diversity of fireflies in many regions of the world because research on the taxonomy of fireflies has not been sufficiently emphasised or adequately funded. As a result, in-depth research has focused on a limited number of species. At the same time, firefly populations are declining across the world, and there is an urgent need for conservation of their habitats. This reflects a decline in the health of the environment and a global trend of increasing biodiversity loss. Governments, local authorities and government agencies need to take measures to preserve the habitats of fireflies and support research, which ultimately provides valuable information to aid in conservation. Protection of the habitats of fireflies contributes to the conservation of many other species of wildlife and a better quality of life for human beings. Fireflies have the potential to be used in education to enhance environmental and conservation awareness. Fireflies have also in recent years become ecotourism icons. Ecotourism needs to be managed sustainably with good, ecologically sound guidelines. Local communities should be involved in the ecotourism industry, be beneficiaries of its economic returns and be involved in conservation.

We, the participants of the Second International Firefly Symposium and Third International Firefly Research Network Meeting recognise and advocate that:

1. Fireflies are a part of our biodiversity heritage and are iconic insects that have been the subject of much investigation in the sciences, an inspiration in the arts and a part of local cultures, folklores and traditions because of their ability to produce light. This ability has also enabled their genes and enzymes to be used in biomedical research for the benefit of mankind.
2. The world's firefly diversity is still poorly understood, and studies on their physiology and behaviour have focused on only a small number of species. Taxonomic research on fireflies has been poorly funded and given insufficient priority, but is greatly needed since it forms the basis for our understanding of their diversity and is crucial for the development of other aspects of firefly research.
3. Fireflies have been a source of ecotourism revenue for many communities in different

parts of the world and have the potential to bring similar benefits to other local communities. Fireflies and their natural habitats also enhance quality of life and contribute to economies through the promotion of aesthetically pleasing landscapes that have greater appeal.
4. Fireflies are indicators of the health of the environment and are declining across the world as a result of degradation and loss of suitable habitat, pollution of river systems, increased use of pesticides in agro-ecosystems and increased light pollution in areas of human habitation. The decline of fireflies is a cause for concern and reflects the global trend of increasing biodiversity loss.
5. Intervention is greatly needed from governments to preserve existing habitats for the conservation of fireflies. The restoration of habitats is not always possible, requires many years and requires high financial inputs.
6. The habitats of fireflies are a refuge for many forms of wildlife including mammals, birds, reptiles, amphibians and numerous species of invertebrates. Conservation of these habitats, therefore, has the potential to conserve a wide range of fauna.

We urge governments, local authorities and government agencies worldwide to take decisive and concrete action at the regional, national and local level to:

1. First and foremost, protect the habitats of fireflies so as to preserve these iconic creatures and other fauna for the enjoyment of future generations.
2. Undertake rehabilitation of degraded firefly habitats to enable gradual recovery of populations.
3. Develop management plans for firefly ecotourism sites that enable them to be managed sustainably and in a manner that is ecologically sound.
4. Promote the involvement of local communities in firefly ecotourism and ensure they are beneficiaries of the economic returns.
5. Ensure local communities are equipped with knowledge of the habitats, life cycle and ecology of fireflies and are fully involved in conservation efforts.
6. Promote education on firefly conservation at the level of schools, and create awareness among the public on the natural history and conservation of fireflies.

We strongly recommend that governments, local authorities, agencies and corporations support the allocation of human and financial resources for:

1. Inventory and documentation of firefly species in highly diverse, under-researched areas such as Asia, Africa and tropical America.
2. Taxonomic research on fireflies at both the morphological and molecular level, which forms a basis for our understanding of the world's poorly-documented firefly fauna.
3. Research on fireflies that provides key information on all aspects of their classification, genetics, biology, ecology, behaviour, physiology, conservation and utilisation in biomedical research.
4. The development and application of low environmental impact techniques that minimise degradation of firefly habitats yet enable the development of infrastructure for the benefit of human communities.
5. Collaboration and the sharing of research findings among researchers, and communication of research findings to government authorities and agencies to aid the cause of firefly conservation.

6. Education on the ecology and conservation of fireflies at the level of schools, local communities and the public, as a tool for inculcating environmental responsibility.

Dated this 5th day of August, 2010, by
the participants of the Second International Firefly Symposium
and Third International Firefly Research Network Meeting
in Subang, Selangor, Malaysia

For further information contact:
The Secretariat, International Firefly Symposium 2010
Entomology Branch, Tropical Forest Biodiversity Centre
Forest Research Institute Malaysia
52109 Kepong, Selangor
Malaysia
Tel. +603-6279 7112
Email: ifs2010@frim.gov.my

Lampyrid 2012. **2**:1–12

Keynote Lecture

Taxonomy – help or hindrance in South East Asia?

Lesley Ballantyne

Charles Sturt University, P.O. Box 58, Wagga Wagga 2678, Australia.
Email: lballantyne@csu.edu.au

ABSTRACT The taxonomic imperative is no less an issue in South East Asia than elsewhere but addressing it is complicated. Early exploration into the natural history of this area resulted in most types and identified material being housed in European museums, with local collections largely consisting of unidentified, or misidentified material. A visit to these European collections is often very difficult, as apart from the financial burden there may be institutional resistance. Access to literature is also a problem as most of the firefly literature from this area was written over 100 years ago. In recent times only two taxonomists have been active in South East Asian firefly taxonomy (in contrast to the activities of several behaviouralists) and their output while welcome barely scratches the surface. Yet at the very beginning of any work on fireflies we need to be able to identify the species reliably, or to determine accurately if it is new. The problems in achieving an adequate taxonomic framework for fireflies in this area are addressed, including how these problems have arisen, why some are particular to the region under consideration, what some of the problems are, and what can be done and is being done to address them. As a result of activities by the symposium organisers, and others in the firefly field, the outlook is decidedly positive.

Keywords: Luciolinae, descriptive taxonomy, South East Asia

INTRODUCTION

The taxonomic imperative regarding fireflies in South East Asia is addressed and the importance of taxonomic studies in this area is stressed.

Taxonomy is defined here as descriptive character based taxonomy but you will find I am not entirely unsympathetic to the ultimate need for other avenues of research as well. My career will be used as an example. I am now 70 years old and I started work on taxonomy of fireflies in 1961 – while I have had a somewhat punctuated career I have lived through the whole of the pendulum swing in regards to attitudes towards the relevance of this science. And there has been a swing of the pendulum as you will see. Taxonomy and the taxonomic endeavour was accepted and supported in the 1960s but along the intervening years for a variety of reasons it lost its foothold. I will concentrate on the Luciolinae for my examples but the situation in the other subfamilies is simply worse and the same comments will apply.

That the taxonomic process is absolutely essential and is the starting point for any investigation of a country's biodiversity is not up for debate. So much has been said by others establishing this point that I refer you to the Zootaxa 2007 special edition (Zhang & Shear 2007).

The particular problems facing S.E. Asia will be outlined, why there are problems and how they are sometimes different from other parts of the world e.g. Australia. How many of the countries in S.E. Asia have an up-to-date list of their firefly fauna (reliably identified) let alone anything else (and this from recent surveys and recent taxonomy, not from a rehash of old catalogues)?

Is there a taxonomic impediment in S.E. Asia? Of course there is, it is a worldwide phenomenon, and not just in relation to fireflies. We are losing species faster than we can describe them and losing habitats. Is taxonomy itself the impediment – a lack of a suitable framework in which to describe species? Partly, but as you will see there are just as many hindrances to actually undertaking taxonomy. However, in outlining the problems I hope to show how we are going about solving them.

OVERVIEW OF TAXONOMY AND ITS IMPORTANCE

Why taxonomy? The year 2010 is the International Year of Biodiversity. Our knowledge of the planet's species is underpinned by taxonomy – it is the cornerstone on which the biological sciences depend.

Then why not taxonomy? The Convention on Biological diversity held in Rio de Janeiro in 1992 had what appeared to be important and far-reaching outcomes. Species are being led to extinction, habitats to destruction and something had to be done. This was 1992. Now in 2010 we still cannot agree as a planet about whether global warming is happening. Taxonomy was clearly vital to the targets set by the Convention in Rio. So why didn't taxonomy get emphasised in reality? Well it did and it didn't. In the USA partnerships like PEET have provided funds and resources to undertake taxonomy and importantly train taxonomists. In Australia the ABRS funds taxonomic projects and recently the unique Bush Blitz program was developed to target both flora and fauna in reserves and National Parks. But here in S.E. Asia it is a different story and some of the older attitudes to taxonomy as a discipline still linger. Money and grants may be sidetracked by the more popular new more interesting areas.

Bocro (2010) argues that the resultant decline in taxonomic endeavour could be partly attributed to the Impact Factor (IF). It is inevitable that taxonomic publications cannot attract a high IF. However the IF has become unfortunately a measure to evaluate scientific work (it was never designed for this outcome). If you are judged by the IF of the journals in which you publish then you will probably not pursue taxonomy and we all lose out and taxonomy as an enterprise itself is being denigrated. There is a perception that because we do not publish in high impact factor journals that our work is not as rigorous or important as that of others who might be doing something more interesting and fashionable. There is no question that our system of valuation of scientific work is flawed – we still refer to Linnaeus after over 200 years – I wonder what his Impact factor might be – at least something for sheer longevity and relevance one hopes? We taxonomists are after all in a non-experimental science and what do we do – we describe things.

Well what is wrong with that? This is the Queen of the Sciences (Jim Lloyd), the science on which all other biological endeavours depend, the starting point for any investigation of biodiversity. And where other methods allow replication of experiments ours is the most testable of all – everything we do from simple examination of specimens, from statements of homology, apomorphy assertions, the hypothesis that is a species, each make explicit predictions about patterns that are testable by further observation.

One of the best ways of getting information out about a species is to do as complete a treatment as possible. After all a name is just a name – we need to know morphological characters, and features of both sexes and of the larvae, where it lives, factors affecting it and so on. For this to mean something, and thus be accountable the specimens examined must be carefully listed and deposited in an established collection. Lloyd and Walker (1967) lamented that biologists who work on a species nevertheless fail to properly identify it. Any work on the biology of a species must be accompanied by properly identified specimens that are lodged in an appropriate collection. The work without vouchers may simply be a waste of time. Because of the uncertainty of our identification of so many firefly species in S.E. Asia we need to have some way of checking subsequently as the taxonomy progresses and voucher specimens are the way.

FIREFLY TAXONOMY IN AUSTRALIA – MY EXPERIENCE

My career has parallels and implications for this part of the world. I started my honours thesis work in 1961 in a Department of Entomology (which is no longer there). There are now no entomology departments in Australia and even the CSIRO Division of Entomology has this year merged with another division to become a mega division 'Ecosystem Science'. Be thankful if you still work in such a department because you will be among others who have some understanding of what it is you are doing and how to support you.

Why did I choose fireflies? I was presented with a list of 20 projects and did not like 1-19 so had to settle by default on 20 which was *Biology of Fireflies*. I had not ever seen a living firefly. I did not start to work on fireflies because of a perceived need. In Australia most people do not know what a firefly is. They visit glow worm caves and think they might be seeing fireflies (glow worms in Australia are larvae of flies). Fireflies in Australia in any numbers occur along the east coast in areas of rain forest that are often national parks, or in the mangroves. Neither area is conducive either to collecting or firefly viewing – one has to beware of the wild pigs, crocodiles and taipans. But what this means is that our firefly fauna is protected. It may not be of interest to many others beside me but – it's protected. It interested me on return home from the Firefly Symposium in Thailand in 2008 to note various participants being interviewed about the alarming decline in firefly numbers worldwide. No one asked me – it was simply not a news item in Australia.

My physical situation in Australia is thus odd for a firefly person. I have lived in Wagga Wagga for close to 40 years – in the arid zone. I am a firefly taxonomist living in an arid zone 500 km from the nearest firefly, in a country that does not appreciate its firefly fauna.

I started to work assembling literature and found on day two that the taxonomic framework had not been updated since our species had first been described mainly in the early 1900s. Ideally, taxonomy is a sort of self-correcting expertise with revisions occurring every 50 years or so. So I had to do some taxonomy to know just what species I was addressing.

I thought it was just going to be a short side trip then I would get back to investigating life history and biology. But there was a difference – I was in a Department of Entomology and taxonomy as a branch of science was welcomed and funded and taught. In 1961 there was no McDermott's catalogue of fireflies of the world (1966), I hadn't even heard of Jim Lloyd. Dare I say I was in the dark about fireflies? In effect I put together a McDermott catalogue of my own. I did not have to translate the literature from a language I'm unfamiliar with as 21 of the 25 species of firefly now known from Australia were described by Australians. I was also able to go and visit the collections and see most of the original types because most of the types and the original collections are in Australia.

THE CHALLENGES IN SOUTH EAST ASIA

How is it different in S.E. Asia? In S.E. Asia, work on fireflies is probably from choice – it might be because everyone knows them, and they are an important part of the economy as a tourist attraction. Thus there is support for that work. However, the challenges faced are tremendous.

You do not own your fauna

Most of the early collections were made by Europeans who then sent their material back to Europe where it was described. The descriptions are in another language (Latin and French mainly and there are few Latin scholars these days). The collections and most importantly the types are in Europe. More S.E. Asian fireflies were described by two Frenchmen – Ernest Olivier and Maurice Pic - than by any other entomologists. The major part, but not all, of these collections is in the Paris Museum. There are problems with actually locating which of the many specimens are actually types. The remaining types are scattered over various European museums. Olivier developed the system of classification for fireflies that was modified somewhat by McDermott but for much of the last 100 years with some exceptions (and those are genera unique to S.E. Asia) we have had to try to use a classification based on European specimens and species. Nowhere is this more obvious than in the Luciolinae where the genus *Luciola* Laporte is based on European representatives and has become a dumping ground for all sorts of different species that simply wouldn't fit anywhere else. Progress is being made in this area.

And what about the museums that once were so supportive of taxonomy? Out of the downturn in the perceived importance of taxonomy there has been a serious downturn in how museums function and how they are funded. If you can afford to visit them then you may have to pay for your bench space. If you want to borrow specimens then you may face an uphill battle. There are fewer and fewer curators to attend to your requests. Museum directors may not always be sympathetic to some of the rigorous procedures taxonomists have to go through and may function more as fund raisers. And a visit to Europe is expensive. It is still easier to get grants for molecular genetics or ecology or conservation than for taxonomy.

However what about museums and the collections of fireflies in S.E. Asia? We should be able to expect that we can identify many species by comparison with what is in existing collections? Without many or any types actually being here in S.E. Asia and as no one has done any recent revisions of the fauna, authoritatively identified collections in S.E. Asia are scarce. One cannot simply work through the material in a collection and assume that the identification is correct. Hence the need to see the types.

Why are types so important? When a species is first described the entomologist has a number of specimens in front of him (usually). These are called syntypes, all being the same species at this stage (or so the taxonomist hopes). Nowadays he/she must select one of these to bear the label of holotype and this is what the species is. No matter if the description does not really seem to match the specimen – the holotype reigns supreme as the indicator of what this species is.

Institutions may not always be sympathetic to your need to visit Paris, for example, even though most of the types may be there. I faced this institutional resistance in Australia, and S.E. Asian taxonomists may face a similar lack of understanding of relevance of taxonomy and lack of comprehension of the discipline. If competing for grants then just taxonomy may not cut it.

The process of taxonomy

What might be involved in the taxonomic process? Let's first address a revision of a genus. Any comprehensive revision should review every species (including synonyms) previously described and locate not only the specimens described originally but as much of what has accumulated since the first revision. If there has not been a previous revision, the same steps apply. Apart from assembling the appropriate literature and translating it if necessary, one must accumulate as many representatives as possible of the species involved from as many collections as will loan their material. Another problem – what if they won't loan, or only loan a few specimens at a time? Visits may be permitted but the problems of prohibitive costs and the institutional resistance remain. There are some countries that will not allow their fauna out of the country – do we need to try to change those attitudes? Again short of actually visiting the country how do we work on the fauna?

One of the major problems facing firefly research in S.E. Asia is the lack of a proper taxonomic framework. We will examine the Luciolinae as this is the subfamily I have worked on. McDermott (1966) listed seven genera and 326 species in the Luciolinae (Table 1).

Of these five genera and two subgenera are exclusively from S.E. Asia and the Australopacific region. These genera and species for the most part existed in the same form in which they were originally described at least 50 years before. Ballantyne and Lambkin (2009) now recognise 19 genera in the Luciolinae of which 17 genera and over 250 species are from S.E. Asia and the Australopacific region. Of these only two genera do not have representation in S.E. Asia. Nine are exclusively from the Australopacific region. Of these genera many have had recent taxonomic updated treatments so we can access up to date descriptions and keys to species of almost all the Luciolinae genera from this region.

Why work on the Luciolinae in this area? At our present state of knowledge S.E. Asia is the area of richest diversity in the world of this subfamily. The Luciolinae include not only those genera known to synchronise and are often important to the economy of a country, but also the most intriguing behaviours and to a morphologist the most interesting array of adaptations in males at the end of their abdomens.

TAXONOMIC ISSUES IN THE LUCIOLINAE

The problem in S.E. Asia in the Luciolinae clearly lies with *Luciola* and it exemplifies the taxonomic problems facing us in S.E. Asia in this subfamily or in others. Below are three specific taxonomic problems that exist in the Luciolinae of this area.

1. The problem with *Luciola* Laporte

The genus *Luciola* is based on *Luciola italica* a European species and thus a European taxonomy. The type is lost, and European researchers do not know just which population there might be the typical one. While many species have been placed in this genus because they did not belong in one of the other better defined genera, the genus is not well defined and is in the opinion of those of us who have to try to work on it a real dumping ground for anything that does not fit elsewhere. However we

Genera[1]	No. species (McDermott 1966)[2]	Current no. species	Taxonomic status[3]
Aquatica Fu et Ballantyne†		5	P
Aquilonia Ballantyne		1	P
Asymmetricata Ballantyne†		2	P
Atyphella Olliff		23	P
Bourgeoisia Olivier	2	2	P
Colophotia Dejean†	10	10	U
Convexa Ballantyne		1	P
Curtos Motschulsky†	7	17	P
Gilvainsula Ballantyne		2	P
Lampyroidea Costa‡	9	9	U
Lloydiella Ballantyne		4	P
Luciola Laporte†	279 (162)	163	P, I
Luciola Laporte‡		?	U, I
Magnalata Ballantyne		3	P
Missimia Ballantyne		1	P
Photuroluciola Pic‡	1	1	P
Pygatyphella Ballantyne		22	P
Pygoluciola Wittmer†	1	6	P
Pteroptyx Olivier†	14	32	P
Pyrophanes Olivier†	5	4	U

[1] Regions indicated by † - S.E. Asia; ‡ - other. Australopacific region genera comprise the remainder.
[2] Not listed in table is one species of *Luciola* subgenus *Hotaria* of the S.E. Asian and Australopacific region. Numbers in parentheses denote Australopacific region species numbers.
[3] P - Published updated taxonomic treatments (descriptions and species keys) available; U - taxonomy updated but not yet published; I - Taxonomic revision in progress.

Table 1 Occurrence and taxonomic status of genera in the Luciolinae in S.E. Asia, the Australopacific region and elsewhere.

are making progress. Ballantyne and Lambkin (2009) and more recently in unpublished data included a number of S.E. Asian "*Luciola*" in their phylogenetic analysis discovered at least 6 distinct clades in *Luciola* all of which should eventually be described as distinct genera. We can access keys to these subdivisions as well (but not to all the species). At present we can assign almost every Luciolinae species from S.E. Asia we see to one of these groups.

But the recent analysis I referred to was hindered by just the same problems as I have already outlined – we simply cannot reliably identify all the species described as *Luciola*. We do not even know just what *Luciola chinensis* is – probably the most populous Asian species, but it like so many other specimens has fallen off its pin in the Linnaean collection in London. It is very probable that a number of species like *Luciola chinensis* L.; *L. praeusta* Kiesenw.; *L. terminalis* Olivier, *L. gorhami* Rits. etc are all just variations on a theme but working out what those variations are and how they are spread across S.E. Asia is an ongoing issue.

2. *Luciola substriata* Gorham, *L. aquatilis* Thancharoen

Luciola substriata Gorham is a firefly that most firefly taxonomists in S.E. Asia will recognise. Larvae are the familiar back swimmers and can be seen at night just below the water surface. However we cannot identify accurately which of the many populations we see in this area is *substriata* sensu stricto. As this species (or variants of it) occur from western India through Thailand to China correct identification is important, and difficult. Already work on specimens indicates that distinctions can only be made by dissections of the genitalia, and that at least two morphologically distinct populations exist in China, one in Taiwan. Gorham's type is no longer recognisable either in Paris or London, but we know the type locality – Bombay, so we should be able to get material from the type locality to help us make a decision. Or should we? India will not allow its fauna to leave its shores. Well may be a visit then? But the same problems of cost apply here and a visit just for one species?

The problem that needs solving gets worse. In the collections of Kasetsart University are a large number of specimens that are morphologically identical with *L. aquatilis* with one exception – they are 50 years old and a different colour. So given that these specimens represent a species that was common on the campus about 50 years ago are we just looking at colour change? In which case does this beautiful and colourful *L. aquatilis* just fade with age? Furthermore given that *L aquatilis* and some of the mainland China specimens we are calling *L. substriata* are also morphologically very similar except for colour – where to now? To further confuse the issue the Chinese specimens appear to start out with this "faded" colour.

3. The problem with *Pteroptyx* Olivier

Species of the genus *Pteroptyx* are tourism in this area. So what if I suggested we might need to change its name? It was based on two species, *testacea* Motsch. and *malaccae* Gorham, both of which were described with deflexed elytral apices and the type species was subsequently considered to be *testacea*. However, the type specimen of *testacea* in Russia does not have deflexed elytral apices – Olivier had misidentified his species. That problem might be overcome but there is another. Ballantyne and Lambkin's ongoing analyses are showing that *Pteroptyx*, *Pyrophanes* and *Colophotia* might all be included in the one genus (you will see in a poster presentation a species which bridges the *Pteroptyx* – *Pyrophanes* gap), and given that *Pyrophanes* was described first this composite genus might just have to bear the name of *Pyrophanes*?

HOW TO PROGRESS?

How then might firefly taxonomy proceed here? What do we need to happen in S.E. Asia? We still cannot identify everything but now we can sort them into species groups based on well defined characters. Frustratingly for some and exciting for others we are in a state of an emerging taxonomy as far as the genus *Luciola* is concerned. Soon we hope these clades will be described as distinct genera with well defined species and the taxonomic process becomes a little easier. However how do we cope with the need for countrywide revisions of the firefly fauna? Work on the firefly fauna of an area requires a taxonomic framework, which for *Luciola* exists only in part. What is needed is a revision from type material of all the fireflies yet known to have been described from S.E. Asia. However, who is going to have the time, inclination or energy to do that? Especially when we all now have some idea of just what that would entail.

Apart from major revisions how should we cope with single species? There is a Thai species that we cannot identify despite an extensive literature search and investigation of the NHM London collection. It is probably new. Descriptions coming out of Asia have little or no phylogeny – there is no background here in that sort of analysis. There has been so little described here and that interferes with undertaking phylogenetic analyses, which themselves slow down the system. Every little publication even if just on one species is a way further along the road than we were before. Species descriptions or redescriptions need to be supported. Thancharoen et al. (2007) in describing a new species of Thai firefly is an example of how to solve that problem.

Does the taxonomist need to be sympathetic to the idea that two groups of people may benefit from his work – the scientific community and on the other hand the man in the street? Should we at least try to produce two sorts of publications – one with lots of coloured pictures as well as the scientific literature? Of course such sorts of sites already exist — Tree of Life, Atlas of Living Australia (but little or no fireflies). But then such books do not earn the academic credibility most require.

THE IMPORTANCE OF COLLABORATION

One of the ways we will solve our taxonomic problems is to cooperate and share information. We need as many specimens from as wide an area as possible – and that is difficult. We need to be able to examine a wide selection of collections and borrow material if we can – so many of these species may now only exist in these old collections. Insects do not respect country boundaries. These are problems faced by all taxonomists in S.E. Asia and are not just restricted to a particular country. Do not be afraid to ask for help – you might be surprised at the outcomes. I could not continue to do what I do without the help of Christine Lambkin who undertakes the phylogenetic analyses for me. Ming-Luen Jeng has always been most open to sharing of information even when there was no outcome other than sharing information with a colleague.

These are examples of some of the collaborations I have undertaken in this area. In each case each author brought to the paper their own expertise, and in each case we were able to solve problems and publish a paper which neither of us could have done by ourselves. In all but one case the authors were widely separated geographically. Ballantyne (the taxonomist) and Lambkin (the cladist) have now examined an ever widening selection of Luciolinae species in Australia New Guinea and S.E. Asia, making inroads into providing a clearer picture of the genera and species within this subfamily (Ballantyne & Lambkin 2000, 2001, 2006, 2009 and in preparation). Ballantyne and Menayah (2000, 2002) described larvae and adults from Malaysia, only possible because of Menayah's field work. Fu and Ballantyne (2006, 2008, 2009, 2010) work on Chinese fireflies with Ballantyne providing the taxonomic expertise and Fu the field skills and technical input. Deheyn and Ballantyne (2009) addressed *Bourgeoisia hypocrita* (Olivier) from Fiji, with Deheyn's field work providing unusual information about the luminescent properties of the female. Thancharoen et al. (2007) solved a taxonomic problem relating to an Asian species that could only have resulted from a cooperative approach. Ballantyne and McLean (1970) cooperated in a taxonomic revision of the genus *Pteroptyx* in the Australian and Pacific region as a precursor to the voyage of the scientific vessel the Alpha Helix to New Guinea, and Ballantyne and Buck (1979) dealt with one of the species taken on that voyage, with Buck providing the field experience.

THE FUTURE

How can we encourage more people to become involved in taxonomy? Firefly taxonomy workshops, such as the one run prior to this Symposium, and a two day workshop held at Kasetsart University after the Thai symposium, presented budding firefly researchers with networking opportunities and developed their taxonomic skills. An upcoming 2011 Bangkok workshop on a similar theme, and several more in planning, are encouraging signs that such programs are perceived as needful. Participants of these workshops probably represent in large part the future of firefly taxonomy in S.E. Asia. These young taxonomists are now a network of like-minded persons who know each other and who understand the firefly taxonomy problems faced in S.E. Asia. The critical need for firefly taxonomists in this region may yet be met, and it is hoped that collectively, we can make a considerable difference in resolving issues encountered in S.E. Asian firefly taxonomy.

REFERENCES

BALLANTYNE, L.A. & BUCK, E. (1979) Taxonomy and behavior of *Luciola* (*Luciola*) *aphrogeneia*, a new surf firefly from Papua New Guinea. *Transactions of the American Entomological Society* **105** 117–137.

BALLANTYNE, L.A. & LAMBKIN, C. (2000) Lampyridae of Australia (Coleoptera: Lampyridae: Luciolinae: Luciolini). *Memoirs of the Queensland Museum* **46** 15–93.

BALLANTYNE, L.A. & LAMBKIN, C. (2001) A new firefly, *Luciola* (*Pygoluciola*) *kinabalua* sp. nov. (Coleoptera: Lampyridae), from Malaysia, with observations on a possible copulation clamp. *Raffles Bulletin of Zoology* **49** 363–377.

BALLANTYNE, L.A. & LAMBKIN, C. (2006) A phylogenetic reassessment of the rare S.E. Asian firefly genus *Pygoluciola* Wittmer (Coleoptera: Lampyridae: Luciolinae). *Raffles Bulletin of Zoology* **54** 31–58.

BALLANTYNE, L.A. & LAMBKIN, C. (2009) Systematics of Indo-Pacific fireflies with a redefinition of *Atyphella* Olliff, Madagascan *Photuroluciola* Pic, and description of seven new genera from the Luciolinae (Coleoptera: Lampyridae). *Zootaxa* **1997** 1–188.

BALLANTYNE, L.A. & MCLEAN, M.R. (1970) Revisional studies on the firefly genus *Pteroptyx* Olivier (Coleoptera: Lampyridae: Luciolinae: Luciolini). *Transactions of the American Entomological Society* **96** 223–305.

BALLANTYNE, L.A. & MENAYAH, R. (2000) Redescription of the synchronous firefly *Pteroptyx tener* Olivier (Coleoptera: Lampyridae) of Kampong Kuantan, Selangor. *Malayan Nature Journal* **54** 323–328.

BALLANTYNE, L.A. & MENAYAH, R. (2002) A description of larvae and redescription of adults of the firefly *Pteroptyx valida* Olivier in Selangor, Malaysia (Coleoptera: Lampyridae: Luciolinae), with notes on Luciolinae larvae. *Raffles Bulletin of Zoology* **50** 101–109.

BOCRO, F. (2010) The study of species in the era of biodiversity: a tale of stupidity. *Diversity* **2** 115–126.

DEHEYN, D. & BALLANTYNE, L.A. (2009) Optical characterization and taxonomic revision of the South Pacific firefly *Bourgeoisia hypocrita* Olivier (Coleoptera: Lampyridae: Luciolinae). *Zootaxa* **2129** 47–62.

FU, X.H. & BALLANTYNE, L.A. (2006) *Luciola leii* sp. nov., a new species of aquatic firefly (Coleoptera: Lampyridae: Luciolinae) from mainland China. *Canadian Entomologist* **138** 339–347.

FU, X.H. & BALLANTYNE, L.A. (2008) Taxonomy and behaviour of lucioline fireflies (Coleoptera: Lampyridae: Luciolinae) with redefinition and new species of *Pygoluciola* Wittmer from mainland China and review of *Luciola* LaPorte. *Zootaxa* **1733** 1–44.

FU, X.H. & BALLANTYNE, L.A. (2009) Larval respiration system and evolution in aquatic fireflies (Coleoptera: Lampyridae: Luciolinae). In: *Bioluminescence in Focus, A Collection of Illuminating Essays* (Ed. by Meyer-Rochow, V.B.), pp. 243–254. India: Research Signpost.

Fu, X.H. & Ballantyne, L.A. (2010) *Aquatica* gen. nov. from mainland China with a description of *A. wuhana* sp. nov. (Coleoptera: Lampyridae: Luciolinae). *Zootaxa* **2530** 1–18.

Lloyd, J.E. & Walker, T.J. (1967) Names are not enough. *Science* **158** 1525.

McDermott, F.A. (1966) Lampyridae Pars. 9. Coleopterorum Catalogus (Junk-Schenkling) (Ed. Sec), 1–149 (Luciolini, 98–118).

Thancharoen, A., Ballantyne, L.A., Branham, M.A. & Jeng, M.L. (2007) Description of *Luciola aquatilis* sp. nov., a new aquatic firefly (Coleoptera: Lampyridae: Luciolinae) from Thailand. *Zootaxa* **1611** 55–62.

Zhang, Z.-Q. & Shear, W.A. (Eds.) (2007) Linnaeus Tercentenary: Progress in Invertebrate Taxonomy. *Zootaxa* **1668**. 766 pp.

Lampyrid 2012. **2**:13–23

Keynote Lecture

The restoration and maintenance of firefly habitats in Japan—case studies from Yokosuka City

Nobuyoshi Ohba

Ohba Firefly Institute, 4-1-12-204 Maborikaigan, Yokosuka City 239-0801, Japan. Email: qgb00523@nifty.ne.jp

ABSTRACT Japanese people have watched fireflies since ancient times, especially during summer months when the attractive Genji-firefly (*Luciola cruciata*) and Heike-firefly (*Luciola lateralis*) fly along the streams and rice fields. The larvae of these fireflies thrive in flowing water of good quality, and may serve as biological indicators of healthy freshwater ecosystems. Urbanisation has encroached upon firefly habitats and reduced firefly populations. It is critical to conserve the habitats in which these fireflies are found. Case studies of successful rehabilitation and creation of firefly habitats in Yokosuka City are presented. Factors contributing to the continuing success of the firefly conservation effort in Japan are discussed. Lessons learnt from the Yokosuka case studies can be applied to future efforts to conserve and restore firefly habitats.

Keywords: Firefly, *Luciola cruciata, Luciola lateralis*, habitat, wetland, restoration, maintenance, conservation, Yokosuka City, Japan

INTRODUCTION

Firefly conservation in Japan is undertaken as a nationwide program administered by local governments and supported by volunteers from the public, non-governmental and non-profit organisations, and private sector companies. The program focuses on habitat restoration and conservation of the riverine *Luciola cruciata* ('Genji' firefly), which is famed for its spectacular summertime light displays, and *Luciola lateralis* ('Heike' firefly), a smaller and more widespread aquatic species found in paddy fields and marshes. Terrestrial firefly species such as *Luciola yaeyamana* ('Yaeyamabotaru'), *Luciola parvula* ('Himebotal'), and also *Rhagophthalumus ohbai* ('Iriomotebotal' or Starworm) are also covered under this program. Aquatic fireflies are sometimes called "Guardians of the Rivers" while terrestrial fireflies are referred to as "Guardians of the Forests." Firefly populations throughout Japan have decreased, and disappeared in some areas, probably due to water pollution by agricultural chemicals, light pollution by street and building lamps, and loss of breeding areas.

The larvae of aquatic fireflies such as *Luciola cruciata* need well-oxygenated, flowing water that is clear and unpolluted. Such conditions, together with a stable gravel bed in the river or stream, and adequate sunshine, support the adhesion and growth of diatoms and other organisms on which small fresh water snails feed. These snails are the main prey of firefly larvae. River banks provide pupation sites, and riverside vegetation needs to be managed by mowing. *Luciola lateralis* needs a constant water level year-round such as is found in rice fields and fallow land. These environments also support the growth of firefly larval food such as tiny aquatic snails and tadpoles. The berms between the rice fields are important pupation sites. The terrestrial firefly species thrive in stable forested environments. The larvae eat land snails, earthworms, and other small organisms. The presence of larvae of *Pyrocoelia fumosa* ('Kromadobotaru') and *Lucidina biplagiata* ('Obabotaru') indicates an excellent and stable wetlands environment that tends to support many other organisms.

This paper presents case studies from Yokosuka city as models of successful

Figure 1 Waterway of Mabori Biological Gardens managed by the Yokosuka City Museum.

firefly rehabilitation and conservation led by the Yokosuka City Museum. The case studies span 30 years and cover over 30 sites in the Miura Peninsula, of which over 20 sites are located in Yokosuka city alone.

CASE STUDIES

Understanding the environmental requirements of these fireflies through long term observations of their behaviour and ecology has been a key to developing rehabilitation and conservation strategies. The main habitat requirements of the aquatic and terrestrial fireflies are good water quality, abundant snail prey, adequate breeding grounds, and unhindered flash communication. The four examples below show how meeting the ecological requirements of the fireflies determine success in rehabilitating their habitats.

Case Study No. 1: An experimental model for firefly rehabilitation

The original site

The 3.8 ha Mabori Biological Garden managed by the Yokosuka City Museum in Kanagawa is a protected area that faces Tokyo Bay. The hill was previously used as an ammunition depot. The secondary-growth forest on the hill is made up of

Figure 2 "Making a home for fireflies" experiment by the Mabori Junior High School, Yokosuka City.

evergreen trees such as *Machilus thunbergii* and *Castanopsis sieboldii*. When the Museum acquired the Garden in 1958, abundant spring water fed the stream that runs through the Garden. Genji and Heiki fireflies populated the stream banks. The construction of a man-made concrete channel in 1975 greatly reduced water flow and modified the water ecosystem, and as a result fireflies subsequently became rare.

Measures taken to restore the site

In 1983, the concrete channel was modified to restore water flow. The channel was widened to 50-100 cm and a shallow weir installed every two metres. A pond was constructed mid-channel to facilitate a more constant flow. Gravel was spread over the channel bed to create a water depth of several cm and to allow plants to grow right up to the banks. A pump was installed 150 m upstream from the pond, to circulate the water at a moderate flow rate that retained a good quantity of leaf litter to support the growth of diatoms. Too slow a flow would cause too much leaf litter to pile up and would hinder water flow in the channel, while too fast a flow would wash away leaf litter. The trees were pruned to allow some sunlight to reach the water channel for the growth of diatoms, and so that thick moss could grow along the banks for the fireflies to lay their eggs. Through trial and error, the water ecosystem was restored by 1986 (Fig. 1). From 2006, the Mabori Junior High School, led by the 'Executive Committee for Creating Firefly Habitats' (formed under a regional alliance of elementary and junior high schools), initiated a programme to rehabilitate and conserve fireflies and to reproduce the programme regionally. Under this programme, a second channel was built that diverted part of the water flow to a manmade marsh (Fig. 2).

The site after restoration

Freshwater snails and other aquatic organisms returned naturally over the years (Water System Study Club 1992) and the fireflies started appearing in June 2008. Fireflies, particularly *L. lateralis* now naturally breed in this modified environment that is maintained by regulating water flow and leaf litter accumulation, and by cultivating moss along the banks.

Case Study No. 2: Habitat rehabilitation through restoration of a river

The original site

Twenty years ago, the Iwato River that flows through Yokosuka City was very polluted, with no fireflies to be seen. Local communities petitioned municipality administrators to clean up the river.

Measures taken to restore the site

In 1984, at the request of the River Bureau of the Ministry of Land, Infrastructure and Transport, the author advised on river restoration in the context of firefly rehabilitation (Water System Study Club 1989). Restoration work began in 1985 and was completed in 1988. The Yokosuka Firefly Club, formed by the local community in 1988, continued restoration efforts that included regular river clean-ups, firefly and aquatic organism surveys, and mowing of vegetation along the river banks (Water System Study Club 1989). This work continues to the present day. Larvae of *L. cruciata* were reintroduced in 1997 and released yearly until 2001.

The site after restoration

Water quality of the Iwato River is now considered good. Large numbers of fireflies emerged after larvae were released at the river. After 2001, the firefly population continued to sustain itself naturally, indicating a stable and healthy reproducing population that now attracts many visitors (Fig. 3). The Iwato River has become the model for rehabilitation of rivers for the return of fireflies in Japan.

Case Study No. 3: Habitat rehabilitation through restoration of a wetland

The original site

The Hikarinooka Wetlands Park, Nobi, Yokosuka City was established in an area

Figure 3 Restored urban river – the case of Iwato River, Yokosuka City.

that was formerly a habitat for *L. cruciata*. The park is surrounded by low hills that encircle the wetlands and rice fields in the valley. A pond collecting rainwater runoff irrigated the surrounding rice fields (Nobi Firefly Study Club 1990). The dragonfly *Minais pruinosa* and firefly *L. cruciata* were found around freshwater sources in the valley. Several hundred *L. lateralis* lived in the marshes, while the terrestrial fireflies *P. fumosa*, *Lucidina biplagiata* ('Obabotal') and *Cyphonocerus ruficolis* ('Munekuriiro-botaru') inhabited the meadow and forest. *Hynobius tokyoensis* (Tokyo salamander) was found in the forest, marshes and flowing water. Its hatchlings develop in the wetlands, returning to the forest as adults. Fresh water snails such as *Semisulcospira libertine, Radix auricularia japonica, Bellamya* (synonym *Cipangopaludina*) *chinensis laeta* and *Austropeplea ollula* lived in the rice fields and streams. A large waterbird, *Anas poecilorhyncha,* and a rare weasel, *Mustela nivalis,* were seen around the pond, and water plants such as the hydrophytes *Penthorum chinense, Saururus chinensis,* and *Nymphoides peltata* grew thickly in the marshes.

Measures taken to restore the site

The wetland area was expanded by reclaiming 30 m of adjacent hillside. Tarp was laid over the reclaimed area to slow water seepage and reduce soil erosion. Water was collected in an underground weir and pumped up to the pond, ensuring continuous water circulation. Care was taken to reproduce the original environment of the wetlands. Three zones were designated for the purpose of park management: 1) a sanctuary zone for restoration of habitats of *L. cruciata* and *H. tokyoensis,* which was thus off limits to visitors, 2) a buffer zone encompassing the wild bird sanctuary, for use by bird watchers under the supervision of guides, and 3) a public zone open to the public. The Park's maintenance was undertaken by members of a nature club, supported by Yokosuka City, and under advice of the author. All the stakeholders agreed on a common long-term target, which was to preserve the park's existing biodiversity and to improve its function as an environmental education center. Aquatic organisms such as *L. cruciata, L. lateralis, P. fumosa* and *H. tokyoensis,* and aquatic plants such as *S. chinensis, N. peltata* and *P. chinense,* were monitored as biological indicators of the progress of restoration in the areas where these organisms would be expected to occur.

The site after restoration

The environmental conditions of the park are now similar to the original conditions before development (Fig. 4), with an accompanying recovery of biodiversity, particularly insects. However, increasing populations of freshwater snails resulted in an increase in nutrient levels that turned the water murky. This was an unforeseen drop in water quality in the gravel-bed brook, which is now silted with mud instead of having the ideal sand-mud ratio. Such conditions are unsuitable for *L. cruciata* larvae, and while more suited for *L. lateralis,* the latter are still only found in small numbers despite large releases. Invasive species such as *Procambarus clarkia* (American crayfish) and *Rana catesbeiana* (bull frog) may be preying on the firefly, while *Procyon lotor* (raccoon) and *Micropterus salmoides* (black bass) may be preying on the fresh water snails that support firefly populations. *Procyon lotor* may also prey on *H. tokyoensis*. Rising populations of *Anas clypeata*, a native species of wild duck, may also be depressing freshwater snail numbers. The effect of these native and invasive species on the fireflies and on *H. tokyoensis* needs further study. Other on-going issues include night-time light pollution from adjacent urban areas, and vegetation growth along riverbanks, which diminishes solar radiation. Night-time light has been reduced by planting tall, thick hedges around the perimeter of the park. Riverside vegetation is pruned periodically to allow sunlight to reach the water. Park maintenance practices were developed through trial and error. The novel concept of a nature club managing a park has spread throughout the country. It is hoped to bring the restored wetlands environment closer to the target of bringing back fireflies by continuing and improving maintenance practices. Public awareness needs to be raised about the objectives of the park, and volunteers recruited.

Case Study No. 4: Creation of an artificial firefly habitat

The original site

This twenty-hectare agricultural park was developed in the Nagai district of Yokosuka City and was launched in March 2005. The scenic Arasaki coast borders the Park. The nearby Nagai fishing port is where fresh seafood is landed. The park has several themes that include a wetland area inhabited by aquatic organisms such as *L. cruciata*, and a gardening area that provides hands-on experience in greenhouse vegetable and strawberry culture and harvesting. Visitors can also experience making bread and handicrafts, and milk cows in the 'Food', 'Craft' and 'Animal Contact' experiences, respectively. The Firefly Museum in the agricultural park (Fig. 5) exhibits the firefly and its habitats.

Figure 4 Hikarigaoka Wetlands Park in Yokosuka City, after rehabilitation.

Measures taken to create the firefly habitat

Construction of an artificial stream about 200 m long, 2-3 m wide and 30 cm deep began in 2004. An underground pump and filtration system was installed to purify and circulate water in the stream at about 20 litres per minute. In 2007, modifications were made to the stream. A river bed was laid, its width narrowed and depth reduced to 5-10 cm. Large rocks were placed to vary the flow of the water, which was also reduced to 1.2 litres per minute. A berm walkway was constructed between rice fields, and a marsh was formed around the waterway. Hydrophytes were planted to slow erosion and purify the water. The upper part of the stream simulated a fast-flowing water environment while the downstream section simulated a pond environment. Both were planted with different species adapted to fast- and slow-flowing water, respectively. After mowing, the riverbanks were planted with *Ludwigia stipulacea* from the Miura Peninsula to preserve this rare plant. Releases of *L. lateralis* were also carried out.

The site after construction

The plants around the man-made wetlands matured, partially covering the water surface and reducing heat from sun exposure. The artificial stream (Fig. 6) has turned out to be an excellent breeding ground for tree frogs, toads and insects, all of which have been found in larger numbers than expected. The Japanese Killi Fish (*Oryzias latipes*), an exotic species, has also been discovered here. *Luciola lateralis* and *L. cruciata* now naturally breed all along the length of the stream and are seen from the middle of May until the end of June. Controlled mowing of plants as part of

Figure 5 Firefly Museum of Nagai Uminote Park, Yokosuka City.

maintenance for the waterway continues to be carried out in April and September.

Other successful firefly conservation efforts

After twenty years of restoration work on the river that flows through Nishihemi Town, *L. cruciata* now breeds naturally there and has repopulated the entire area. The local community identified suitable habitats for restoration work. The river banks were also raised and reinforced to prevent floods.

The local communities in five towns in Oppama district have been actively creating firefly habitats along the shores of Tokyo Bay. One of these habitats is found near Hashimizu Elementary School. Four elementary schools have been enlisted in the activity of 'Making a Home for the Firefly'. All the sites are within walking distance from a railway station.

Nobi district, which is a large wetland area high in biodiversity and having a natural abundance of fireflies, was selected as the base for 'Making a Home for the Firefly' activities that have now been extended to Kinugasa Park, Yamanaka-cho and Akiya. These areas form a network of wetlands around the whole Miura Peninsula. Similar activities are being organised in Miura City, Hayama Machi, Zushi City, and Kamakura City, in the Miura Peninsula. There has been good cooperation from all the suburbs in the city.

Figure 6 Creation of an artificial firefly habitat in the Nagai Uminote Park.

Conservation of the terrestrial fireflies *R. ohbai* ('Iriomotebotaru'), *L. yaeyamana* and *L. owadai* ('Kumejimabotaru') has been carried out in Kumejima (Ohba 1996), Iriomotejima, Ishigakijima and Okinawa Island, as part of a biodiversity research programme for the Ministry of the Environment. Conservation efforts and research on other terrestrial firefly species such as *L. parvula* are ongoing.

DISCUSSION

Two factors have contributed to the success of creating and restoring firefly habitats in Yokosuka City:

1) A vast knowledge base on the ecological requirements of fireflies.
The availability of much research material on fireflies, the existence of an excellent model for firefly rehabilitation and the availability of expertise have been the basis for successful restoration of firefly habitats. The Yokosuka City Museum has been a research base for fireflies for 50 years and has accumulated a large amount of relevant research material. The Mabori Biological Garden, managed by the Museum for the past 50 years, remains the best model for the creation of an artificial firefly habitat and successful introduction of fireflies. The author has vast experience in firefly rehabilitation, having assisted various government and public interest groups in restoring and maintaining over 30 firefly habitats in the Miura Peninsula. The cornerstone in rehabilitation is creating or restoring the habitat so that it can support the natural breeding of freshwater snails on which the fireflies feed, thus favouring natural breeding of fireflies. When these conditions are met, monitored introductions of fireflies from other areas may be carried out to augment existing firefly populations.

2) Support from the government and the participation of schools and NGOs.
Generous financial support from the government at the national and district levels have enabled firefly rehabilitation work to progress well in Japan. Schools, NGOs and communities in districts across Japan have contributed volunteer manpower that has driven firefly restoration efforts. Their work has been coordinated by the "Japan Firefly Society," a nationwide organisation based in Aomori City in northern Japan that has been carrying out firefly rehabilitation since 1967.

CONCLUSION

The considerable effort that has been devoted to rehabilitating firefly habitats and restoring firefly populations in Yokosuka City has met with much success. United in the common objective of conserving fireflies, strong bonds of cooperation have been forged amongst all participating parties. Using fireflies as a 'mascot' or charismatic species in conservation efforts has helped protect and restore areas in which many other plant and animal species now flourish.

However, firefly habitats and the biodiversity contained within them continue to be threatened by pollution, timber harvesting, receding water levels and invasive alien species. Some of the areas where the fireflies are found are privately owned, and loss of these habitats to agricultural activities and urbanisation is inevitable. It is hoped that a balance will be sought between such activities and ecological needs. Cultivated rice fields and forests that are managed for sustaining firefly populations are such examples. Fireflies have proven to be good indicators of the success of restoring habitats. Monitoring by counting the number of firefly flashes is straightforward and can be carried out by volunteers.

The author hopes that other countries will also endeavour to preserve their firefly habitats for future generations to enjoy.

REFERENCES

Nobi Firefly Study Club. (1990) Ecological research on the firefly and analysis of its habitats, Nobi area, Yokosuka City, Kanagawa Prefecture. *Science Report of the Yokosuka City Museum* **38** 47-60.

Ohba, N. (1996) Ecological study of the firefly, *Luciola owadai* and conservation of its habitat. *Insect and Nature* **31** 17-21.

Water System Study Club. (1989) Recovery of the urban water system, Iwato River, Yokosuka, Kanagawa Prefecture. *Science Report of the Yokosuka City Museum* **37** 9-28.

Water System Study Club. (1992) Quantitative character analysis of urban water system environment in the Miura Peninsula, Kanagawa Prefecture. *Science Report of the Yokosuka City Museum* **40** 11-19.

Lampyrid 2012. **2**:24–30

First record of the firefly genus *Pygoluciola* Wittmer (Coleoptera: Lampyridae: Luciolinae) in Thailand

Wandee Wattanachaiyingcharoen* & Sorasak Nak-eiam

Department of Biology, Faculty of Science, Naresuan University, Phitsanulok, 65000, Thailand.
**Email: wandeew@nu.ac.th*

ABSTRACT Fireflies of the genus *Pygoluciola* Wittmer are rare in South-East Asia. Previously they were known only from the Malaysian states of Sabah and Sarawak, Mindanao Island in the Philippines and a single species in mainland China. Here a first record from Thailand is presented. This taxon conforms morphologically to an expanded view of the genus lacking the pronounced abdominal modifications.

Keywords: *Pygoluciola*, Lampyridae, Luciolinae, firefly genitalia, Thailand

INTRODUCTION

The rare South East Asian firefly genus *Pygoluciola* Wittmer was described for one species, *P. stylifer*, from Borneo (Wittmer 1939). The generic name derives from the pronounced terminal modifications of the abdomen in the male, where the median area of both ventrite 7 and tergite 8 are prolonged and incline towards each other. *Pygoluciola* was listed as a subgenus of *Luciola* by McDermott (1966), and redescribed by Ballantyne (1968) from four species including two that were new, and the type species *stylifer* Wittmer. The terminal abdominal modifications and genital complex were figured for the first time. Ballantyne (1968) considered the genital complex of *Pygoluciola* distinctive, and suggested the elevation to generic level again. Earlier phylogenetic analyses on a reduced number of Luciolinae species did not further resolve the position of *Pygoluciola* (Ballantyne & Lambkin 2000, 2001). Subsequently analyses involving increasing numbers of both Luciolinae species and morphological characters indicated that *Pygoluciola* should be returned to generic status (Ballantyne & Lambkin 2006, 2009). This was formalised by Ballantyne and Lambkin (2006) with an expanded generic definition. Ballantyne (2008) described another new species from the Philippines and reported on the seven species included in the genus viz. *P. guigliae* (Ballantyne), *P. hamulata* (Olivier), *P. kinabalua* (Ballantyne & Lambkin), *P. stylifer* Wittmer and *P. wittmeri* (Ballantyne) from the Malaysian states of Sabah and Sarawak, and *P.*

satoi Ballantyne from Mindanao in the Philippines. Here we group and characterise these as 'island species,' which conform in abdominal and genital characteristics to *Pygoluciola* sensu Wittmer (1939). These species, to this point considered rare, exhibit the following modifications of the terminal abdominal segments in the male described above. The aedeagus is longer than wide with the median lobe apex acute and usually much shorter than the pale apparently fleshy lateral lobes (Ballantyne 1968; Ballantyne & Lambkin 2006). The aedeagal sheath is symmetrical with a long narrow sheath sternite, which expands well behind the lateral tergite articulations (Ballantyne & Lambkin 2006: Figs. 15-17).

Fu and Ballantyne (2008) presented an expanded generic description of the genus and described *Pygoluciola qingyu* from Yuan An, China, which we characterise here as the sole 'mainland species.' *Pygoluciola qingyu* lacks those modifications on the terminal abdominal segments characteristic of the island species, but the male genitalia features are typical of described *Pygoluciola*. Ballantyne and Lambkin's (2009) phylogenetic analysis supported the expanded generic definition and the inclusion of *qingyu* in *Pygoluciola*.

Pygoluciola in either form has not previously been recorded from Thailand. Theraphat (1969) surveyed fireflies in Bangkok and Samut Prakarn provinces in the central region, finding *Luciola* (now *Asymmetricata*) *circumdata* Motsch. and one *Pteroptyx* species, and an unidentified *Luciola* sp. In 1989, Lloyd et al. reported flashing behaviour observations on four *Luciola* species from Bangkok namely, *L. japonica* (Thunb.), *L. cingulata* Olivier, *L.* nr *carinata* and *L. circumdata* Motsch. The investigations on fireflies in Thailand undertaken under the HM Queen Sirikit's royal initiative firefly project from 1997-2000 reported the presence of ten genera of fireflies from 35 provinces including only three Luciolinae viz. *Luciola* Laporte, *Pteroptyx* Olivier, *Pyrophanes* Olivier (Firefly Project under HM Queen Sirikit's Initiative Committee 2001). Hutacharern et al. (2007) listed 15 species of fireflies found in Thailand including the Lucioline species *Luciola brahmina* Bourgeois, *L. cruciata* Motschulsky, *L.*(now *Asymmetricata*) *ovalis* Hope, *L. substriata* Gorham, *Pteroptyx malaccae* (Gorham), *P. valida* Olivier, and *Pyrophanes indica* Motschulsky. Thancharoen (2001) surveyed the highland region of Khao Yai National Park in the northeastern region and lowland area of Nakhon Pathom, Samut Prakarn and Samut Songkram provinces in the central region, and recorded five genera including *Luciola* Laporte, and *Pteroptyx* Olivier. Sommitr (2004) reported 4 genera of firefly from the Botanic garden of Chiang Mai province in the northern region, including *Luciola* Laporte, and *Pyrophanes* Olivier.

In April, 2008, five adult male fireflies from the lower montane forest of Kamphaeng Phet province at 1,300 metres above mean sea level were collected by sweep net and preserved in 70% ethanol. This collection was identified as *Pygoluciola* species by using Ballantyne and Lambkin's (2009) keys, and confirmed by Ballantyne in 2010 (personal communication).

In this paper we briefly describe the new record of the genus *Pygoluciola* found in Thailand with figures of the genital complex of the unidentified species. A full description will follow upon confirmation of the identity of this potentially new species.

METHODS

Genitalia of five males were dissected following Ballantyne and McLean (1970). Both male genitalia and aedeagal sheaths were examined using the key to genera (Ballantyne & Lambkin 2009). Taxonomic characters are based on Fu and Ballantyne (2008).

RESULTS

Description of *Pygoluciola* sp.

Specimens examined Thailand, Kamphaeng Phet Province, 1,300 m, April 23, 2008, S. Nak-eiam. 5 males (Department of Biology, Naresuan University, Thailand).

Diagnosis: Distinguished from all *Pygoluciola* species except *P. qingyu* by lacking an elongate, narrow median posterior projection (MPP) that curves dorsally and a median prolongation of tergite 8 that curves ventrally. Differing from *P. qingyu* in its colouration (*qingyu* has a pinkish-yellow pronotum with median dark markings, and dark brown to black elytra) and in the length of the median lobe of the aedeagus, which in *qingyu* is about three quarters as long as the lateral lobes, while the median lobe of *Pygoluciola* sp. is less than half as long as the lateral lobes.

Male: Figs. 1 (1 & 2). Body elongate, slender; 9.0 – 10.5 mm long; 2.9 – 3.0 times as long as wide.

Pronotum: 1.35 – 1.6 mm long, 2.6 – 3.0 mm wide, width/length = 1.8 – 1.9; 1/6.5 – 1/7 as long as whole body length; dorsal surface of pronotum mainly fairly smooth and flat; punctures small, shallow, fairly inconspicuous. Median anterior margin rounded and slightly projecting beyond obtusely rounded anterolateral corners. Lateral pronotal margins posteriorly divergent and slightly convergent at posterior half. Hypomera flat in posterior one third. Posterolateral corners acute, projecting beyond median posterior margin and delimited from the median posterior margin by shallow emarginations. Posterior margin slightly, narrowly depressed, and emarginated in the posteromedian area.

Head: Posterior one third of head retracted into prothoracic cavity at rest; vertex moderately depressed; greatest head width 1.9 – 2.2 mm; smallest interocular width 0.32 – 0.42 mm; antennal sockets separated by less than antennal socket width. Mouthparts well developed; apical segment of labial palpi laterally flat, with the longest inner edge bearing a number of slender finger-like projections, varying to 3-7 teeth on both sides. Antenna length less than twice maximum head width. Antennae eleven segmented, filiform; all flagellar segments elongate, slender, simple; apical flagellar segments (FS) never conspicuously shortened or clavate, not produced laterally or flattened.

***Elytra*:** 6.8 – 8.0 mm long. Elytra sub-parallel sided without deflexed apices. Semitransparent brown with the lateral margins quite pale yellow colour; pale creamy sutural margins in apical half with tiny dark brown apices. Punctuation not linear, very dark brown; four interstitial lines, delimited at their sides by punctures; Epipleura only slightly expanded along its length. Epipleura and sutural ridges extending to rounded apices.
***Abdomen*:** Fig. 1 (2). All ventrites very dark brown. White pale light organ in V (ventrite) 6 and V7. V7 light organ restricted to mesal two thirds, not attaining lateral margins, posterior margin shallowly emarginated. Median posterior projection of V7 very short with rounded apex and not upturned. Tergite 7 wider than long, laterally depressed. Tergite 8 slightly wider than long, medially not produced or ventrally inclined. Lateral margins sub-parallel-sided in posterior half and anteriorly convergent in anterior half. Posterior margin of tergite 8 broadly rounded; anterior margins bifurcate into two elongate narrow projections.
***Legs*:** Tibia straight. Femora not swollen or curved. Legs with coxae, trochanters and femora yellow. Tibiae and tarsi dark brown.
***Aedeagal sheath*:** Figs. 1 (3-5). About 3.5 times as long as wide. Anterior three quarters of sheath sternite elongate, symmetrical and narrow, expanding in posterior one quarter to a symmetrical heart shaped area, and terminating along its posterior margin in paired, hairy, rounded projections. Tergite of sheath projecting at sides with long, lateral, well defined sclerotised arms directed posteriorly and articulating with sternite of sheath at its anterior one quarter. Sheath tergite with anterior margin broadly emarginate and posterior margin truncate.
***Aedeagus*:** Figs. 1 (6 & 7). Median lobe about half as long as lateral lobes, apex narrow, round. Lateral lobes separated for almost all of their length dorsally. Basal third of lateral lobes well sclerotised. Anterior margin of lateral lobes produced asymmetrically; apical two thirds of lateral lobes pale fleshy, not sclerotised. Apices of lateral lobes slightly out-turned, bearing hairs along most of their lateral margins, especially more dense in basal third. Small short pointed projections along inner dorsal margin at the middle.

DISCUSSION

Within the Luciolinae there are pronounced modifications of the male terminal abdominal ventrite and tergite, which are often characteristic of a particular grouping (Ballantyne 1987; Ballantyne & Lambkin 2009). The upturned inclination of the median posterior projection of ventrite 7, and the downturned inclination of the median posterior margin of tergite 8 that are also narrowly produced were the most obvious features of the genus *Pygoluciola* (Wittmer 1939; Ballantyne 1968; 2008; Ballantyne & Lambkin 2001, 2006). However, in 2008, Fu and Ballantyne reported a new species, *P. qingyu* from mainland China that lacks the modifications of the male terminal abdominal segments, but has very similar genitalic features in

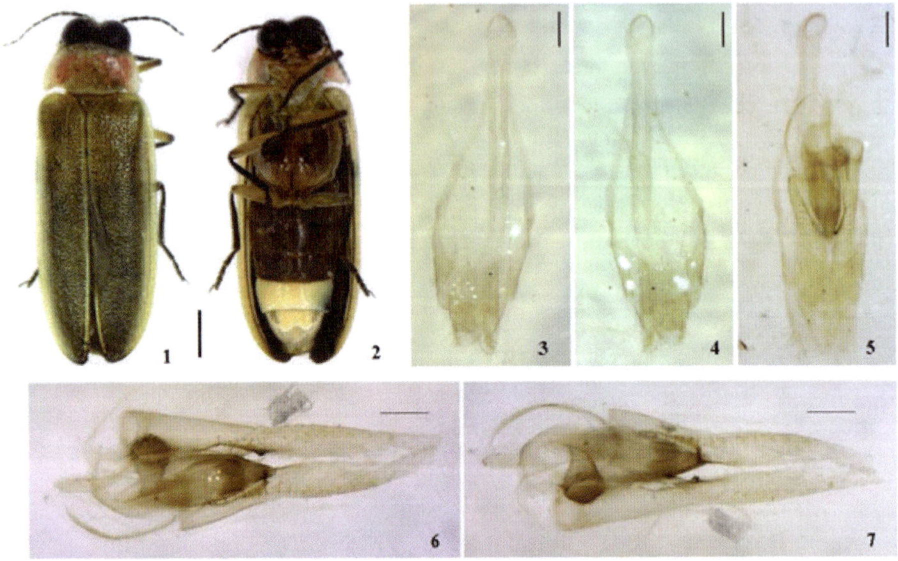

Figure 1 *Pygoluciola* species adult male habitus: (1) dorsal and (2) ventral; (3) aedeagal sheath ventral; (4) aedeagal sheath dorsal; (5) aedeagal sheath with aedeagus dorsal; (6) aedeagus dorsal; (7) aedeagus ventral. The scale line in (1) and (2) = 1 mm; the scale lines in (3) - (7) = 0.2 mm.

both male and female to those of *Pygoluciola* already described. They expanded the definition of *Pygoluciola* Wittmer, and keys to genera of Luciolinae were modified to include features of the aedeagal sheath and male genitalia (Fu & Ballantyne 2008; Ballantyne & Lambkin 2009). This new record of a species of *Pygoluciola* found in Thailand lacks those modifications of the male terminal abdominal segments while conforming to *Pygoluciola* in male genitalia.

The long separation of the Borneo islands and Mindanao island from the mainland of the Asian continent may have led to the separation of the mainland and island species. The unmodified male terminal abdominal segments appear to characterise the mainland species (mainland Chinese species, *P. qingyu,* and the Thai *Pygoluciola* sp.), which while sharing the same feature of the unmodified male terminal abdominal segments, differ in body colour, morphology of the V7 light organ and especially the length of the lateral lobes of the aedeagus.

ACKNOWLEDGEMENTS

This study was financially supported by the TRF/BIOTEC Special Program for Biodiversity Research and Training grant BRT R_152077, the Diversity and Ecology of Fireflies in Thailand Project under HM Queen Sirikit's initiative and the Faculty of Science, Naresuan University. The facilities used in this research were provided by the Department of Biology, Naresuan University. Lesley Ballantyne provided useful comments on drafts of this manuscript.

REFERENCES

Ballantyne, L.A. (1968) Revisional studies of Australian and Indomalayan Luciolini (Coleoptera: Lampyridae: Luciolinae). *University of Queensland Papers*, Department of Entomology, University of Queensland Press vol II, no. 6, 105-139.

Ballantyne, L.A. (1987) Lucioline morphology, taxonomy and behavior: a reappraisal (Coleoptera, Lampyridae). *Transactions of the American Entomological Society* **113** 171-188.

Ballantyne, L.A. (2008) *Pygoluciola satoi*, a new species of the rare S. E. Asian firefly genus *Pygoluciola* Wittmer (Coleoptera: Lampyridae: Luciolinae) from the Philippines. *Raffles Bulletin of Zoology* **56** 1-9.

Ballantyne, L.A. & Lambkin, C. (2000) The Lampyridae of Australia (Coleoptera: Lampyridae: Luciolinae: Luciolini). *Memoirs of the Queensland Museum* **46** 15-94.

Ballantyne, L.A. & Lambkin, C. (2001) A new firefly, *Luciola* (*Pygoluciola*) *kinabalua*, new species (Coleoptera: Lampyridae), from Malaysia, with observations on a possible copulation clamp. *Raffles Bulletin of Zoology* **49** 363-377.

Ballantyne, L.A. & Lambkin, C. (2006) A phylogenetic reassessment of the rare S. E. Asian firefly genus *Pygoluciola* Wittmer (Coleoptera: Lampyridae: Luciolinae). *Raffles Bulletin of Zoology* **54** 21-48.

Ballantyne, L.A. & Lambkin, C. (2009) Systematics of Indo-Pacific fireflies with a redefinition *Atyphella* Olliff, Madagascan *Photuroluciola* Pic, and description of seven new genera from the Luciolinae (Coleoptera: Lampyridae). *Zootaxa* **1997** 1-188.

Ballantyne, L.A. & McLean, M.R. (1970) Revisional studies on the firefly genus *Pteroptyx* Olivier (Coleoptera: Lampyridae: Luciolinae: Luciolini). *Transactions of the American Entomological Society* **96** 223-305.

Firefly Project under HM Queen Sirikit's Initiative Committee (2001) *Fireflies in Thailand.* Project Report. Queen Sirikit Botanic Garden, Office of the Prime Minister. (in Thai).

Fu, X.H. & Ballantyne, L.A. (2008) Taxonomy and behaviour of lucioline fireflies (Coleoptera: Lampyridae: Luciolinae) with redefinition and new species of *Pygoluciola* Wittmer from mainland China and review of *Luciola* LaPorte. *Zootaxa* **1733** 1-44.

Hutacharern, C., Tubtim, N. & Dokmai, C. (2007) *Checklists of Insects and Mites in Thailand* (Lampyridae, 30). Department of National Parks, Wildlife and Plant Conservation, Bangkok.

Lloyd, J.E., Wing, S.R. & Hongtrakul, T. (1989) Flash behavior and ecology of

Thai *Luciola* fireflies (Coleoptera: Lampyridae). *Florida Entomologist* **72** 80-85.
McDermott, F.A. (1966) Lampyridae (Luciolini, 98-118). *Coleopterorum Catalogus Supplementa*. Volume 9 (Second Edition).
Sommitr, W. (2004) *Population Ecology of the Firefly, Pyrocoelia praetexta Olivier in Dok Toh Botanic Garden*. M.Sc. Thesis. Chiang Mai University (in Thai).
Thancharoen, A. (2001) *Study on Diversity of Firefly Populations in Highland and Lowland Habitats*. M.Sc. Thesis. Mahidol University.
Theraphat, T. (1969) *Study on Life Cycle, Development and Light Producing Organs in the Firefly, Luciola circumdata Mots*. M.Sc. Thesis. Chulalongkorn University (in Thai).
Wittmer, W. (1939) Fauna Buruana. Coleoptera 7. Beitrag zur Kenntnis der Indo-malayischen Malacodermata. *Treubia* **17** 21-24 (Lampyridae).

Lampyrid 2012. **2**:31-36

Impact of artificial light on the distribution of the common European glow-worm, *Lampyris noctiluca* (Coleoptera: Lampyridae)

STEFAN INEICHEN* & BEAT RÜTTIMANN

Institute of Natural Resource Sciences, Zurich University of Applied Sciences, Grüental, 8820 Waedenswil, Switzerland.
**Email: s.ineichen@bluewin.ch*

ABSTRACT The influence of artificial light on fireflies is often discussed. Specialists agree that the normal mating behaviour of fireflies is disturbed by artificial night lighting. The effects of artificial light on the behaviour of the common European glow-worm, *Lampyris noctiluca*, are not well understood. Males of *L. noctiluca*, which can fly but do not glow, are attracted to glowing, flightless, larviform adult females, and are easily caught by simple light-emitting diode (LED) traps. LED traps were placed in illuminated and dark areas (under and between street lights, respectively) in a residential quarter of the village of Biberstein in Switzerland. On nights when the street lights were on, the LED traps attracted males only in dark places, while traps under the street lights stayed empty. In contrast, on nights when the street lights were switched off, males were trapped both between and under the lights. This indicates that artificial night light interferes with the ability of males to locate females, or that males avoid illuminated areas. Night lighting, however, had no apparent influence on the spatial distribution of glowing females during the last six summers (2004–2009). Females displayed both under the lamps and in the dark areas bordering the path surveyed. As females normally do not move from their display sites, those displaying under street lamps may die without mating. Artificial night lighting may, therefore, cut 'cheese holes' in the mating landscape of glow-worms.

Keywords: Artificial light, light pollution, mating behaviour, glow-worm, *Lampyris noctiluca*

INTRODUCTION

Artificial night light, which has become common in populated areas, may have negative impacts on fireflies. The mating activity of fireflies that are active at twilight is triggered by diminishing light intensity and can be delayed by artificial light, which may also decrease the efficiency of firefly bioluminescent emissions as mating signals (Lloyd 2006). Firefly specialists agree that light pollution may be one of the reasons for a decline in firefly populations around the world, but few studies are available to support this hypothesis. A laboratory study of *Luciola aquatilis* in Thailand noted that artificial light conditions prolonged courtship and dorsal mounting behavior, thus lengthening mating time (Thancharoen et al. 2009). No field studies have been published on the influence of artificial light on firefly behaviour.

The common European glow-worm, *Lampyris noctiluca*, which is the subject of this study, is a widespread species with a range extending from Portugal and Britain in the west, across Europe and Asia to China in the east (Tyler 2002). During summertime in Switzerland, and especially in June and July, the flightless larviform adult females begin to glow soon after dusk. They often display on low vegetation in open spaces, for example along waysides, forest edges and grasslands. With their bioluminescent signal, the females attract males, which fly in obstacle-free zones in search of a mate (Ineichen 2004). Soon after mating, females stop glowing. The same individual females of *L. noctiluca* have been found glowing night after night under street lamps and in other illuminated sites for up to 17 nights (Ursula Moor, pers. comm. in 2009). The objective of this study was to determine the effect of artificial night light on the ability *L. noctiluca* females to attract males, using simulated female glows emitted by LED (light-emitting diode) traps.

Figure 1 Map of the residential area in Biberstein, Switzerland where *Lampyris noctiluca* was sampled. LED traps targeting males were placed along 210 m of a pedestrian lane illuminated at night with street lamps. Yellow spots mark each trap. Red circles show the locations of lamps. A small creek (not visible in the map) flows on the right side of the lane and empties into a pond indicated by the white arrow. The red box marks the area where the distribution of females was examined in 2007.

METHODS

The study was conducted along a 210 m section of a pedestrian lane (midpoint of lane 47°24'59"N, 8°05'07"E; Fig. 1) flanked by a creek, in the residential quarter of the village of Biberstein in Switzerland. Vegetation bordering the path largely comprised grasses and small shrubs in a more or less open space.

The positioning of five high pressure sodium street lamps along the lane resulted in alternating illuminated and dark areas beneath and between the lamps, respectively, which were used to investigate the effects of illuminance on the glow-worm. Male glow-worms were sampled using LED traps on 14 fair-weather nights between 14 June and 7 July 2010. For three of the sampling nights (28 June, 1 and 4 July), the street lamps were switched off. The LED traps, which are designed to simulate female glow behavior, have been established to be an effective method of sampling *L. noctiluca* males (Ineichen 2004; J. Tyler, pers. comm. in 2007). One LED trap was placed beneath each of the five street lamps (five illuminated areas), between the street lamps and at both ends of the lane (six dark areas) as shown in Fig. 2. The traps were operated between 10.00–11.15 pm, at the end of civil twilight, which occurred between 10.06–10.09 pm during the study period (sunsets took place between 9.26–9.29 pm). Trapped males were collected and put in a box and released at the end of each nightly sample. Illuminance at each trap site was measured using a luxmeter (Voltacraft® MS-1300).

Figure 2 A LED trap (lower left) in an illuminated area beneath a street lamp. Another trap (not visible) is placed in the dark area between this and the next street lamp in the distance.

The mean number of males caught per night was calculated for each trap, from the total catch over the 11 days when the lamps were turned on and, separately, for the three days when the lamps were switched off. Means per trap were standardised for differences between periods when lamps were on and periods when lamps were off by dividing them by the overall mean for all traps when lamps were on and off, respectively. Differences between standardised mean catches on lit and unlit nights were compared within traps placed under and between street lamps, and were tested for significance using paired t-tests. The results were examined in relation to the spatial distribution of glowing females in this area in 2007, using data provided by Ursula Moor.

RESULTS

A total of 73 males were trapped throughout the 11 nights with illumination and all were exclusively in the dark areas (between lamps). Sampling on the three nights with no illumination yielded 58 males in total from traps both under and between lamps. One third of the combined total for all nights were captured between 10.30–10.45 pm. Standardised mean catches under lamps on lit and unlit nights differed significantly (paired t-test, $T = -0.56$, $P = 0.005$), with all catches occurring only on unlit nights (Fig. 3). In non-illuminated areas between lamps, standardised mean catches on lit and unlit nights were also significantly different (paired t-test, $T = 5.14$, $P < 0.005$), with catches on lit nights exceeding catches on unlit nights (Fig. 3). On lit nights, illuminance beneath street lamps ranged from 46–64 lux, and at dark areas measured between 0.1–0.4 lux. During the three nights with no illumination, maximum illuminance was 0.4 lux.

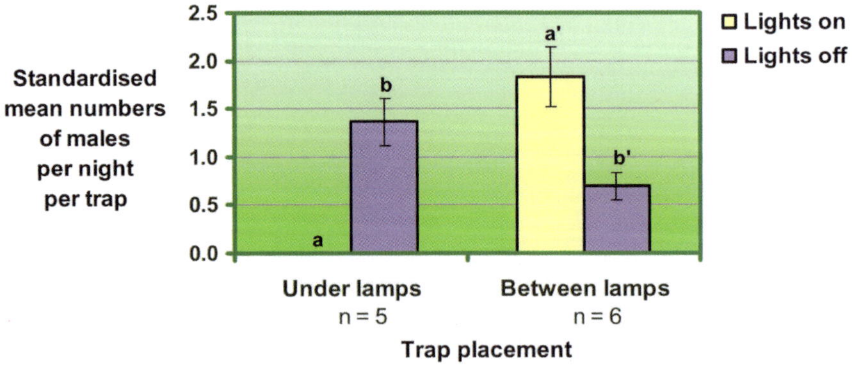

Figure 3 Standardised mean numbers (± S.E.) of *L. noctiluca* males captured per night per trap under and between street lamps when lamps were switched on or switched off. Means with the same letter within trap placement categories were not significantly different (paired t-test, $P \leq 0.005$).

Records of the distribution of females in that area in 2007 (data courtesy of Ursula Moor) revealed some clustering around the pond and along some stretches of the path next to the stream (Fig. 4). However this clustering appeared independent of the street lamps, i.e., females were just as often found displaying near street lamps as in the dark areas between lamps. Further, in our own study, we counted a total of 31 females displaying within 5 m of the street lamps and 76 females in the dark areas outside the 5 m radius of the cones of light, although actual numbers in dark areas may be higher than what was counted because females stop glowing soon after mating, which in dark areas could take place earlier.

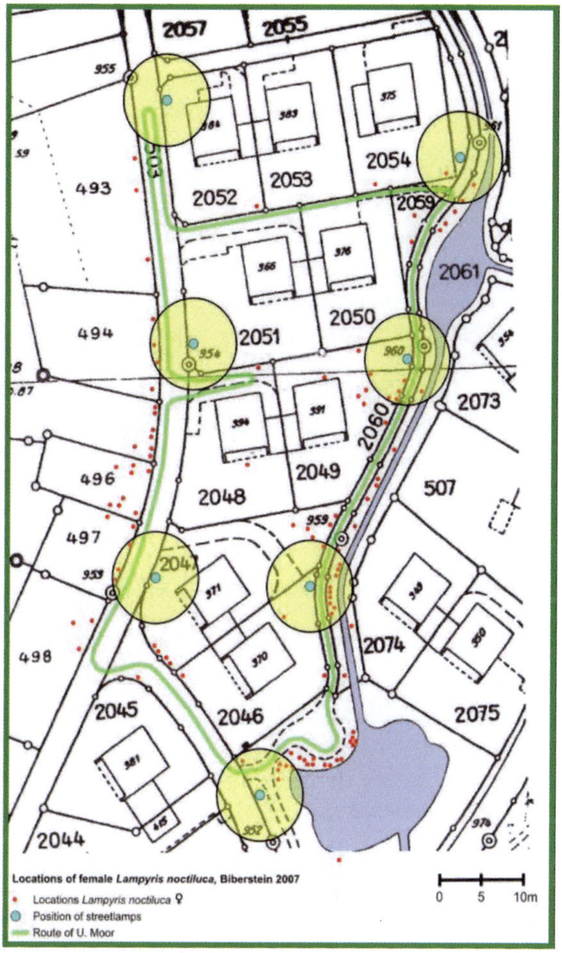

Figure 4 Distribution of *L. noctiluca* females along the streets and pedestrian pathway of a residential area of Birbenstein, Switzerland in the summer of 2007. Distribution map kindly provided by Ursula Moor.

DISCUSSION

The results clearly indicate that artificial night light interferes with the ability of *L. noctiluca* males to locate females, or that the males avoid illuminated areas. This is evident from the absence of any males in traps under lamps but their presence between lamps on nights when the lamps were switched on, which contrasted sharply with their presence in traps both under and between lamps on nights when the lamps were switched off. Within the cones of light produced by the lamps, illuminance was as much as 550 times greater than in areas only a few meters away.

The high number of females displaying in illuminated areas suggests that female glow-worms do not base their selection of display sites on whether or not that area is artificially illuminated at night. In their final instar, they select their pupating and displaying sites during the day, preferring vegetation that borders on open spaces (personal observations). The vegetation provides refuge in the day and a perch from which they can display and attract males after dusk. However, as adult females rarely move from their original selected site (Ursula Moor, pers. comm.), it is likely that females displaying beneath street lamps die without having mated.

Artificial night light is, therefore, anticipated to carve "cheese holes" in the mating landscape of glow-worms. Whether these sinks where females die under street lamps without mating endanger the local population remains to be seen and warrants further study. Although in some areas artificial night light may lead to the extinction of common glow-worms, in the well-planned residential quarter of Biberstein the population of *L. noctiluca* does not appear to be negatively affected at this time.

ACKNOWLEDGEMENTS

We are very grateful to Ursula Moor who generously shared data gathered from hundreds of her night walks recording female glow-worms. We also thank John Tyler for correcting the text and Raphaël De Cock and Ingo Rieger for reviewing an early draft of this paper.

REFERENCES

Ineichen, S. (2004) Zur Raumnutzung von Larven, Weibchen und Männchen des Grossen Glühwürmchens *Lampyris noctiluca. Mitteilungen der Entomologischen Gesellschaft Basel* **53** 111-122.

Lloyd, J.E. (2006) Stray Light, Fireflies, and Fireflyers. In: *Ecological Consequences of Artificial Night Lighting* (Ed. by Rich, C. & Longcore, T.), pp. 345-364. Washington: Island Press.

Thancharoen, A., Baimai, V. & Branham, M.A. (2009) *Luciola aquatilis*, a new species of aquatic firefly from Thailand. In: *Diversity and Conservation of Fireflies. Proceedings of the International Symposium on Diversity and Conservation of Fireflies, Queen Sirikit Botanic Garden, Chiang Mai, Thailand, 26-30 August 2008* (Ed. by Napompeth, B.), pp. 26-29. Chiang Mai: Queen Sirikit Botanic Garden.

Tyler, J. (2002) *The Glow-worm*. Sevenoaks: Lakeside Printing Ltd. 75 pp.

Lampyrid 2012. **2**:37-47

The distribution, life cycle and flash behaviour of the firefly *Pyrocoelia praetexta* Olivier (Coleoptera: Lampyridae) in the Alishan area of Taiwan†

Jen-Zon Ho[1*], Jing-Han Hu[1], En-Cheng Yang[2], Chia-Hsiung Wu[2] & Ping-Shih Yang[2]

[1] *Taiwan Endemic Species Research Institute, C.O.A. No. 1 Minsheng E. Road, Jiji, Nantou County 552, Taiwan.*
[2] *Department of Entomology, National Taiwan University. No. 1, Sec. 4, Roosevelt Road, Taipei 106, Taiwan.*

**Email: jenzonho@tesri.gov.tw*

ABSTRACT The firefly *Pyrocoelia praetexta* Olivier is abundant in the Alishan area in Taiwan. The vertical extent of the Alishan area ranges from 400–1500 m above sea level. Sixteen species of terrestrial snails, including eight new records (*Cyclophorus taivanus, Cyclophorus formosensis, Ptychopoma wilsoni, Coniglobus nux, Coniglobus arisana, Aegista subchinensis, Aegista* (*Plectotropis*) *impexa*, and *Pancala batanica pancala*), have been identified as prey of this firefly. From September to December late instar larvae of fireflies were found preparing for pupation and many flashing males were observed, with the peak being from October to December, depending on the elevation. Two basic life cycles of *P. praetexta* were observed. Type I is univoltine with overwintering eggs and type II is bivoltine with overwintering larvae. It was found that at higher elevations (1200-1500 m) the adults emerged earlier than adults at the lower elevation of 400 m. The average body length of an adult firefly at the higher mountain elevations was significantly larger than that of an adult firefly in the lower mountain elevations ($p<0.01$). The imaginal period lasted from September to October. Maps of the vertical distribution of *P. praetexta* were plotted using a geographical information system and showed that distribution is related to elevation and local flora of the Alishan area. The male photic emission was 525.2–590.8 nm, with the emitted peak wavelength λ_{max} at 547.44 nm. Conservation problems and firefly watching activities in the Alishan area are also discussed.

Keywords: *Pyrocoelia praetexta*, life cycle, occurrence, distribution, habitat, illumination, Alishan

† *This paper was submitted to the symposium proceedings but was not presented orally as the author was unable to attend the symposium.*

INTRODUCTION

There are over 60 species of *Pyrocoelia* fireflies recorded in the world, with most species found in the Asian region in countries such as China, Japan, Thailand, Vietnam, Malaysia, India, South Korea, Taiwan, and the Philippines (Furukawa 2010). Taiwan is home to five species of *Pyrocoelia*, i.e., *praetexta*, *analis* Fabricius, *sanguiniventer* Olivier, *formosana* Olivier and *prolongata* Jeng and Lai (Miwa 1931; Jeng et al. 1999). *Pyrocoelia* fireflies are noted to be habitat-specific with low dispersal capacity (Ohba et al. 1995). Adult fireflies are somewhat flat, with the head hidden under the pronotum and marked with two transparent spots on the margin of the pronotum. The brachypterous females do not fly.

Pyrocoelia analis is well studied, with information available on distribution, habitat requirements (Ho 2004; Wu & Perng 2007), oviposition behaviour (Ho et al. 2002a), factors affecting egg hatching and larval survival (Ho & Huang 2004), pupation and emergence (Ho et al. 2002b) and its life cycle in the laboratory (Ho et al. 2003). Information on the other species is limited to a study on the biological characteristics of *P. pectoralis* (Wang et al. 2007) and reflex-bleeding (Fu et al. 2006), and a laboratory-based life cycle description for *P. praetexta* (Chen et al. 1998). The life cycle of *P. praetexta* has never been studied in the field. This species, commonly seen from autumn to winter in the mountainous areas within its range from Dalin and Chiahsien to Sheliao, is the largest and brightest of the Taiwanese *Pyrocoelia* (Chen 1999; Ho & Ju 2002). Populations are in decline (Yang 1998) due to habitat conversion to agriculture and other development projects. One such area where the population decline has been noted is the Alishan National Scenic Area in Chiayi County. Alishan is known globally for its scenic vistas and natural attractions including firefly-watching. This mountainous region covers about 41,520 ha and ranges in climate from tropical to subtropical to temperate as elevation increases (200–2,600 m). In the protected area the high biodiversity is well maintained (Yang 1999; Perng 2001; Ho 2006) but agricultural activities around it have negatively affected the firefly population. Here we describe the life cycle of *P. praetexta* in the field and its distribution in Alishan, in the context of its conservation.

METHODS

Study area

A total of 17 sampling sites were selected from the four administrative areas (Alishan Township, Chuchi Township, Meishan Township, and Fan Road Township) in Alishan. Historical annual averages (1981–2010) for temperature, relative humidity and rainfall are 11.2°C, 86.2% and 3932 mm, respectively, for Alishan (2010, Central Weather Bureau: www.cwb.gov.tw/eng/index.html). Of the 17 sites, 10 were surveyed monthly while the rest were occasionally surveyed. The ten sampling sites selected were forested areas with minimal light pollution that were easily accessible, have potential for eco-tourism development, and in a range of elevations between

400–1285 m a.s.l. The remaining seven sites that received occasional visits were located in areas outside the 400–1285 m a.s.l. altitudinal range, i.e., at 250 m and up to 2400 m a.s.l.

Sampling procedure

Between March 2005 and February 2006, a one km transect line was set up along the roads at each of the 10 sampling sites. The sites were surveyed monthly by a four man team. Surveys started at 7:00 p.m. and ended at about 9:00 p.m., and took about a week to complete. In the event of rain, which happened on one occasion, the survey was postponed to the following day. The total number of males within 2 m of each side of the line was counted. After the counts were completed and recorded, a 35 cm diameter butterfly net on a 110 cm long pole was used to capture flashing males on the wing for 30 minutes. Captured males were temporarily placed in a transparent sealed bag (34 x 47 cm). Larvae, which could be detected by their faint glow, were collected together with snails they were feeding on, and placed in a transparent container. At the end of the collecting period, captured males were measured (length and width). About five adults and larvae and their snail prey were brought back to the laboratory for rearing while the others were released. Several additional adults were also brought back to the laboratory for spectral analysis.

Laboratory studies

In the laboratory, the larvae were provided with the terrestrial snail, *Bradybaena similaris,* as food (after Ho et al. 2003) and reared till adult. Emergent females and males were paired for mating and oviposition. Male light spectra were measured using the S2000 Fiber Optic Spectrometer, with the optical fiber attached to the males' light organ. Five males were used. Five males each from five selected sites were also measured to investigate size variation in relation to elevation. All specimens were deposited at Taiwan Endemic Species Research Institute.

Data analysis

The mean body length of *P. praetexta* at different elevations was analysed using one-way ANOVA (SPSS-GLM-Univariate V.13), followed by Tukey's honestly significant difference to test for significant differences between the means.

RESULTS

Distribution and occurrence

P. praetexta larvae and/or adults were collected from the lowest site sampled (250 m a.s.l.) all the way up to the second highest site (1380 m a.s.l.) but not at the uppermost site (2400 m a.s.l., Alishan National Forest Recreation Area). The altitudinal distribution of *P. praetexta* was estimated to be from 250 to 1380 m a.s.l. based on the survey (Table 1). Males from Chukou Ecological Education Park (250 m a.s.l.) and Naoliaoxi Ecological Engineering Park (400 m a.s.l.) were found as late as December, with a peak numbers in November. The other sites within the 400–1104 m a.s.l. range also yielded greater catches in November while sites at higher elevations yielded peak catches in October.

Life cycle and spectral emission

As the life cycle is not easily observed in the field, some experiments were conducted in the laboratory. Using measurements of larval body length, we compared field individuals with laboratory-reared field-collected larvae. Most laboratory-reared individuals pupated before September but a few continued to develop through the larval stage until the next year. Therefore, we conclude that *P. praetexta* has both univoltine and bivoltine life cycles, with females of normal and larger sizes, respectively. The mated female oviposits under damp leaves or loose, fairly moist soil. The eggs then overwinter and hatch the following year between March and May. The hatched larvae search for snail prey at night and can be found glowing brightly on the forest floor, particularly in valleys and gullies where moisture is higher. Glowing first instar larvae tend to aggregate in the field from late April to mid May. The larvae develop quickly during spring and summer and are mature by September. Around that time they start to search for a place to pupate. They may also overwinter as larvae, under leaves or stones, until the following spring.

Based on the field observations, we found 14 species of arboreal and terrestrial

#	Sampling sites	Elev.	Location	2005										2006		Forest
		(m)		M	A	M	J	J	A	S	O	N	D	J	F	Type
1	Chukou Ecol. Edu. Park	250	23°30'36.0"N 120°48'17.7"E	-	-	-	-	-	-	-	0	2	2	0	-	A
2	Naoliaoxi Ecol. Eng. Park	400	23°24'41.4"N 120°09'58.0"E	0	0	0	0	0	0	0	1	30	5	0	0	B
3	Feng Shan Firefly Tr.	800	23°33'45.8"N 120°18'44.9"E	0	0	0	0	0	0	0	0	12	0	0	0	B
4	Yuantanxi Ecol. Park	862	23°30'32.3"N 120°12'44.9"E	0	0	0	0	0	0	0	0	2	0	0	0	A
5	Tefuye	885	24°10'56.9"N 120°32'59.9"E	-	-	-	-	-	-	0	-	2	1	-	-	E
6	Yuantanxi Rest Stop	888	23°30'50.7"N 120°12'26.8"E	0	0	0	0	0	0	0	0	1	0	0	0	A
7	Dingbenzi Firefly Tr.	940	23°28'08.6"N 120°14'04.2"E	0	0	0	0	0	0	0	1	5	0	0	0	A
8	Tapang	1100	23°25'17.0"N 120°17'31.8"E	-	-	-	-	-	-	-	1	-	0	-	0	E
9	Ruolan Resort	1104	23°31'22.3"N 120°13'55.3"E	0	0	0	0	0	0	0	0	7	0	0	0	A
10	Aiyuzi Ecol. Park	1130	23°30'46.7"N 120°18'18.0"E	-	-	-	-	-	-	0	1	4	0	-	-	F
11	Lanpo	1170	23°32'13.7"N 120°14'05.3"E	0	0	0	0	0	0	0	2	0	0	0	0	E
12	Shuisheliao	1176	23°29'16.5"N 120°12'39.6"E	0	0	0	0	0	0	0	11	5	0	0	0	F
13	Changshan Trail	1220	23°30'51.1"N 120°15'01.5"E	0	0	0	0	0	0	0	10	0	0	0	0	F
14	Fenqihu South Tourist Tr.	1285	23°29'02.1"N 120°14'47.9"E	0	0	0	0	0	0	1	10	5	0	0	0	C
15	Temple Trail of Fenqihu	1369	23°29'14.3"N 120°14'38.0"E	-	-	-	-	-	-	-	3	0	-	0	-	C
16	Fenqihu North Tourist Tr.	1380	23°29'20.9"N 120°14'52.9"E	-	-	-	-	-	-	1	2	-	-	-	0	C
17	Alishan Nat'l. Forest	2400	20°25'09.2"N 120°20'47.9"E	-	-	-	-	-	-	0	-	0	0	-	-	C

Table 1 Number of male fireflies captured in a survey of sites from 250 to 2400 m a.s.l. in Alishan. Dashes (-) indicate months where no survey was conducted. Forest type: A, bamboo-broadleaf; B, broadleaf; C, conifer plantation; D, orchard; E, bamboo-conifer; F, orchard-conifer.

Figure 1 The hypothetical life cycle *Pyrocoelia praetexta* in Alishan: Type I is univoltine with overwintering eggs while Type II is bivoltine with overwintering larvae.

snails hunted by the larvae, six of which were previously reported, i.e. *Bradybaena similaris, Leptopoma taivanum, Videna haenseli, Aegista mackensii, Nesiohelix swinhoei* and *Achatina fulica* (Chen et al. 1998) and eight of which were new to the list of *P. praetexta* larval food prey: *Cyclophorus formosensis, Cyclophorus taivanus, Ptychopoma wilsoni, Coniglobus nux, Coniglobus arisana, Aegista subchinensis, Aegista* (*Plectotropis*) *impexa*, and *Pancala batanica pancala* (Fig. 2). The larvae avoided snail species that were highly sticky and also larger species of snail, e.g., the African giant snail, *Achatina fulica,* which is 4 cm in length.

With regard to effects of elevation, mean body length of males collected from Naoliaoxi Ecological Engineering Park (400 m a.s.l.) was significantly shorter than that of males from sites at higher elevations ($F=6.09$, d.f.=4, $p< 0.01$, n=5; Fig. 3).

The emission spectrum of *P. praetexta* males measured in the laboratory showed peak emission wavelengths (λ_{max}) ranging from 544.69–547.44 nm (n=10).

DISCUSSION

Distribution and occurrence

The firefly was collected from Fenqihu North Tourist Trail (1380 m a.s.l.), but not from the Alishan National Forest Recreational Area (2400 m a.s.l.). Therefore, the upper limit at which *P. pratexta* occurs in Alishan lies between 1380 and 2400 m a.s.l. Additional sampling at sites between these two elevations is needed to better define the upper altitudinal limit of this firefly. Also, this firefly may occur at elevations below 250 m a.s.l. as sampling was done no lower than Chukou Ecological Park (250 m a.s.l.) where *P. praetexta* was found. Additional sampling at sites below this elevation is needed to determine the lower limits at which this firefly occurs.

It is interesting that numbers of *P. praetexta* appeared to peak about a month earlier at higher elevations than in sites at lower elevations. This phenomenon is unusual. Herbivorous insects, particularly those with an annual life cycle in which adult emergence is promoted by temperature or temperature combined with photoperiod, are often progressively delayed in emergence with increasing altitude (Hodkinson 2005). We suggest that in contrast with species that emerge in warmer seasons, fall-emerging *P. praetexta* possibly need low temperatures and short-day photoperiod to promote its eclosion, causing the earlier emergence with increasing altitude. Further extensive sampling is needed to determine the correlation between peak occurrence and elevation and the factors influencing the seasonality of this firefly.

There is some overlap in the distribution of *P. praetexta* and *P. analis*, both of which were found at Chukou Ecological Park. *Pyrocoelia analis* is common at elevations below 250 m a.s.l., in moist grassland, riverbeds and bamboo plantations in Western Taiwan (Ho 2004). The males of *P. analis* can be seen at night, from September to December. Their green-yellow flashes can be seen as they fly around searching for a mate (Chen et al. 1998; Ho 2006).

Figure 2 Terrestrial snail prey of *Pyrocoelia praetexta* larvae in Alishan. A-H: *Coniglobus nux, Leptopoma taivanum, Cyclophorus formosensis, Cyclophorus taivanus, Acusta tourannensis, Aegista mackensii, Achatina fulica* and *Nesiohelix swinhoei*, respectively.

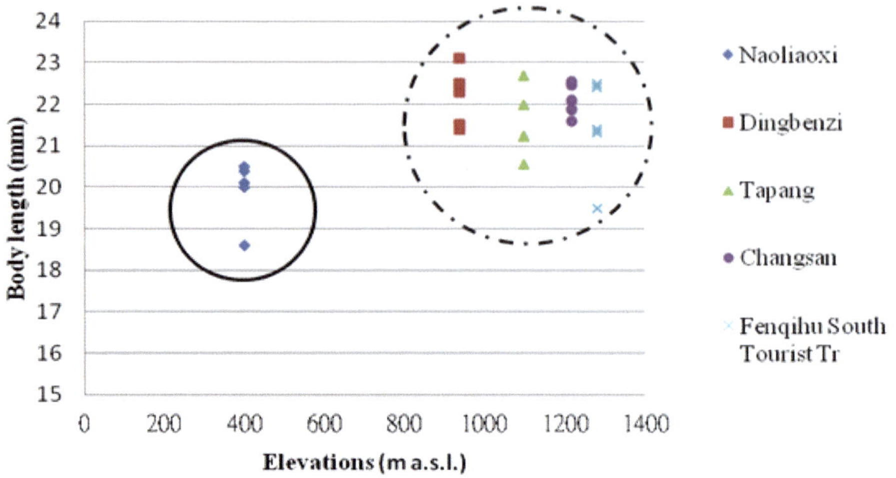

Figure 3 Body length of *Pyrocoelia praetexta* males from five sites of different elevations at Alishan. Naoliaoxi (400 m a.s.l.) Dingbenzi (1050 m a.s.l.), Tapang (1100 m a.s.l.), Changsan (1220 m a.s.l.) and Fenqihu South Tourist Tr (1285 m a.s.l.).

Life cycle and spectral emission

The apparent occurrence of both univoltine (Type I) and bivoltine (Type II) life cycles for *P. pratexta* in this study is a phenomenon that would benefit from further investigation. This firefly species exhibited a univoltine life cycle in a laboratory study in northern Taiwan (Chen et al. 1998), which has temperatures similar to that of the lowest elevation sampled in Alishan. In that study, the female and male life cycles were 365 and 335 days, respectively (Chen et al. 1998). In insects, because the number of generations per year decreases when latitude and altitude increases (Tauber et al. 1986), we expect that *P. pratexta* may be univoltine at lower elevations with the incidence of bivoltinism increasing at higher elevations at Alishan.

The significantly smaller males at 400 m a.s.l. as compared with those at higher elevations is likely due to temperature differences. At lower temperatures, insects are often larger because they develop slowly (Danks 2007).

The emission spectrum of *P. praetexta* (λ_{max}: 544.7–547.4 nm) falls within the range exhibited by *P. pectoralis* (526.5–593.9 nm) found in the Hubei Province of China, but *P. pectoralis* has a higher λ_{max} (548.6 nm). The two species are very similar in their external morphology so they may be distinguished by their λ_{max} values. The emission spectrum of *P. praetexta* is also similar to that of three European firefly species, *Lampyris noctiluca*, *Phosphaenus hemipterus,* and *Lamprohiza splendidula* (De Cock 2004) but is lower than that of 22 firefly species (Asian, 16 spp.; Jamaican, 1 sp.; American, 4 spp.) studied by Seliger et al. (1964). In the latter study, the spectra ranged between 552 and 585 nm.

Prospects for conservation of *P. praetexta*

Firefly eco-tourism in Alishan has increased substantially in recent years and the government recognises it is important to conserve the firefly to ensure the continued growth of this industry. As the habitats of *P. praetexta* overlap extensively with agricultural areas in Alishan, the key to successful conservation of the firefly is to maintain the integrity of its habitats using sustainable agricultural practices. Recent efforts by the government to promote organic methods in cultivation of the high-value crops grown here, e.g., tea, persimmon and coffee, have reduced the use of pesticides, herbicides and chemical fertilisers. The decrease in environmental pollution from agricultural runoffs has had a positive impact on firefly populations, and *P. praetexta* no longer faces an immediate danger of extinction.

Initiatives by volunteers and local residents to maintain the firefly's habitat and conduct annual monitoring of firefly populations also bode well for the future of firefly eco-tourism in Alishan. The development of a formal conservation plan for *P. praetexta* could draw from the case study of *Pyrocoelia rufa* conservation in Japan where community-led and government-supported initiatives have had great success. Also known as the Aki firefly, *P. rufa* is similar in appearance to *P. praetexta* and its very bright flash displays can be seen in autumn. *Pyrocoelia rufa* has been strictly protected by the local government since 1966 when it was designated a Memorial Natural Creature in Nagasaki County.

In Alishan, firefly watching activities are normally held in spring and early summer. However, the surveys in the present study have found some sites where *P. praetexta* light-displays continue to autumn, thus there is potential for extending the firefly watching season in Alishan. This would allow for the development of new eco-tourism projects supported by Alishan National Scenic Area Administration and Chiayi Forest District Office. An action plan is needed to ensure sustainable eco-tourism development that will benefit the local community, while conserving the firefly.

ACKNOWLEDGEMENTS

The authors are grateful to Mr. Xiao-Yu Tang, the director of ESRI, for his encouragement. We also gratefully acknowledge the assistance from Mr. Tzong-Jeng Lin, Mrs. Shu-Lin Chang and Miss Nuo-Hua Tsia with firefly breeding and field investigations. Snail specimens were identified by Dr Chi-Li Tsai who is a taxonomist of Mollusca in TESRI.

REFERENCES

Chen, S.C., Chen, Z.C. & Yang S.C. (1998) Life history of *Pyrocoelia praetexta*. *Bulletin of Ilan Technology* **1** 53-58 (in Chinese).

Chen, T.R. (1999) *An Ecological Guide to the Taiwanese Fireflies*. Taipei: Field Image Press. 191 pp. [in Chinese].

Danks, H.V. (2007) The elements of seasonal adaptation in insects. *Canadian Entomologist* **139** 1-44.

De Cock, R. (2004) Larval and adult emission spectra of bioluminescence in three European fireflies species. *Photochemistry and Photobiology* **79** 339-342.

Fu, X.H., Ohba, N., Meyer-Rochow, V.B., Wang, Y.Y. & Lei, C.L. (2006) Reflex-bleeding in the firefly *Pyrocoelia pectoralis* (Coleoptera: Lampyridae): morphological basis and possible function. *Coleopterists Bulletin* **60** 207-215.

Furukawa, Y.H. (2010) *Firefly Encyclopedia.* http://tokyo-hotaru.com/jiten/firefly.html.

Ho, J.Z. (2004) Occurred fluctuation, distribution and habitat characters of the firefly, *Pyrocoelia analis. Formosan Entomologist* **24** 117-128 (in Chinese).

Ho, J.Z. (2006) *Surveying for firefly resources and publication of guide book for firefly watching activity in Chia-yi mountain area.* Firefly Conservation Society of R.O.C. 100 pp (in Chinese).

Ho, J.Z. & Huang, S.W. (2004) Effects of temperature and egg size on egg duration, hatching rate, and starvation tolerance of first instar larvae of the firefly, *Pyrocoelia analis. Formosan Entomologist* **24** 305-311 (in Chinese).

Ho, J.Z., Huang, S.W. & Su, T.H. (2002a) Egg production and ovipositional behavior of the firefly, *Pyrocoelia analis* (Fabricius). *Formosan Entomologist* **22** 43-51 (in Chinese).

Ho, J.Z., Huang, S.W. & Su, T.H. (2002b) Using a digital video system to study pupation and emergence of the firefly *Pyrocoelia analis* (Fabricius). *Endemic Species Research* **22** 43-51 (in Chinese).

Ho, J.Z. & Ju, J.C. (2002) *Guide Maps of Taiwanese Firefly Watching.* Taichung: Morning Star Publishing Inc. 325pp (in Chinese).

Ho, J.Z., Su, T.H. & Huang, S.W. (2003) Rearing methods and life cycle of *Pyrocoelia analis* (Coleoptera: Lampyridae). *BioFormosa* **38** 79-87 (in Chinese).

Hodkinson, I.D. (2005) Terrestrial insects along elevation gradients: species and community responses to altitude. *Biological Reviews* **80** 489-513.

Jeng, M.L., Lai, J., Yang, P.S. & Sato, M. (1999) On the validity of the generic name *Pyrocoelia* Gorham (Coleoptera, Lampyridae, Lampyrinae), with a review of Taiwanese species. *Japanese Journal of Systematic Entomology* **5** 347-362.

Miwa, Y. (1931) *A Systematic Catalogue of Formosan Coleoptera.* Formosa: Department of Agriculture, Government Research Institute.

Ohba, N., Goto, Y. & Kawashima, I. (1995) Competitive behavior of mating males in the firefly, *Pyrocoelia abdominalis* (Coleoptera: Lampyridae). *Science Report of the Yokosuka City Museum* **45** 39-44 (in Japanese).

Perng, L.C. (2001) *Wild Plant Resources of Jaiyi County.* Nantou County: Taiwan Endemic Species Research Institute Publication. 224 pp (in Chinese).

Seliger, H.H., Buck, J.B., Fastie, W.G. & McElroy, W.D. (1964) The spectral distribution of firefly light. *Journal of General Physiology* **48** 95-10.

Tauber, M.J., Tauber, C.A. & Masaki, S. (1986) *Seasonal Adaptation of Insects.*

New York: Oxford University Press. 411 pp.

WANG, Y.Y., FU, X.H., LEI, C.L., JENG, M.L. & OHBA, N.Y. (2007) Biological characteristics of the terrestrial firefly *Pyrocoelia pectoralis* (Coleoptera: Lampyridae). *Coleopterists Bulletin* **61** 85-93.

WU, W.C. & PERNG, J.J. (2007) Habitat environmental factors and population fluctuations of the firefly, *Pyrocoelia analis* (Coleoptera: Lampyridae). *Formosan Entomologist* **27** 31-45 (in Chinese).

YANG, P.S. (1998) *Fireflies.* Taipei: Nature Conservation Association of R.O.C. 83 pp (in Chinese).

YANG, J.C. (1999) *Wild Animal Resources of Jaiyi County.* Nantou County: Taiwan Endemic Species Research Institute Publication. 224 pp.

Lampyrid 2012. **2**:48-71

Fireflies in the snow: Observations on two early-season arboreal fireflies *Ellychnia corrusca* and *Pyractomena borealis*

LYNN FAUST

Emory River Land Company, 11828 Couch Mill Road, Knoxville, TN USA 37932. Email: TNLFaust@gmail.com

ABSTRACT Field studies on two cold-hardy sympatric firefly species *Ellychnia corrusca* complex (Linnaeus) and *Pyractomena borealis* (Randall) were conducted from 2008 to 2011, in East Tennessee in the southeastern United States. Their behaviours, predation events, escape strategies, calendar and degree-day parameters for life events are presented and regional differences discussed. Adults of *E. corrusca* and last instar larvae of *P. borealis* become active in February when snow and below-freezing temperatures are common. They gather in the furrows on the warmer, sunlit south sides of still-leafless preferred trees, the majority of which are large diameter oaks (*Quercus* sp.) for *E. corrusca*, hickories (*Carya* sp.) for *P. borealis* and tulip poplars (*Liriodendron tulipifera*) for both firefly species. *Ellychnia corrusca*, lanternless and diurnal, abruptly appear low on colony trees in late winter (Jan-Feb), orienting to the sun and vertically to one another upon initial emergence. A 4- to 6-week quiescent period follows this emergence during which time they remain on their colony tree slowly patrolling or resting head down in the bark furrows. Flight, dispersal and mating occur in mid-March to April. Diurnal *P. borealis* larvae climb from cold winter ground up preferred trees seeking sunny, protected pupation sites. Frequent moving is common in Jan-Feb as larvae select ideal arboreal sites. Pupation takes place in March with protandrous males attaching, pupating and eclosing before most females. Adult males use two mating strategies. Initially day active, they seek out and tend both larval and pupal females for up to two weeks and attempt copulation as soon as the guarded virgin female ecloses, bypassing courtship and flashing. In the nocturnal courtship display, *P. borealis* males flash in flight every ca. 2 s in April on fair-weather nights above 10-12°C, high in the forest canopy. Intense male competition, repeated mating and prolonged (1- to 8-day) copulation is common.

Keywords: *Pyractomena borealis, Ellychnia corrusca*, daytime lightning bug, mate guarding, pupal tending, firefly, preferred trees, arboreal, aerial pupation

INTRODUCTION

In early to mid-February (late winter) with few insects evident, two sympatric firefly species become active in the bare, cold, often snow-covered forests of the Great Tennessee (TN) Valley, the Great Smoky Mountains and the Cumberland Mountains of East TN in the southeastern United States (Fig. 1). Adults of the *Ellychnia corrusca* L. complex (Fender 1970), diurnal and lanternless, emerge and share many of the same trees as the last instar larvae of *Pyractomena borealis* Randall that climb from the forest floor to seek arboreal pupation sites on the southern aspect of the trunks of those trees (Fig. 2).

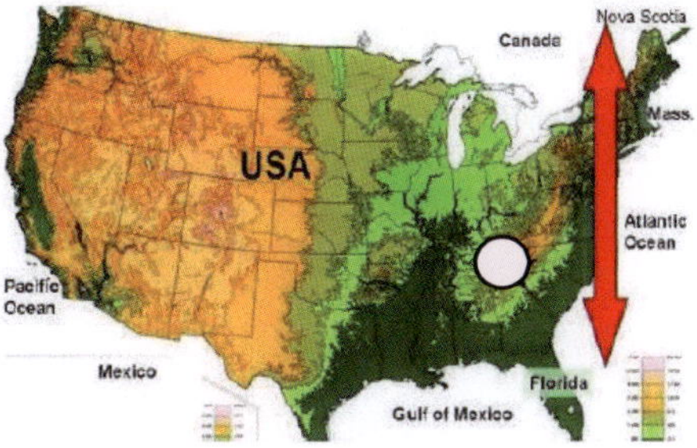

Figure 1 East TN sites (white circle: TN). North-south range (red arrow) of both *Pyractomena borealis* & *Ellychnia corrusca* complex.

Both species are widespread in the eastern half of the US with the latitudinal range extending south from southeastern Canada to north central Florida (FL), a distance of more than 2200 km (Green 1957; Fender 1970; Bousquet 1991; Lloyd 1997; Rooney & Lewis 2000) (Fig. 1). Initial natural history observations were recorded for both species almost a century ago by Williams (1917). *Pyractomena borealis* preimaginal larval descriptions are reported by Archangelsky and Branham

Figure 2 *Ellychnia corrusca* (A) and *Pyractomena borealis* (B). Note the pale elytral margin of the latter.

in 1998. Buschman (1984) describes the transformation of the prepupal pygypod (moveable tail organ) to a permanent anchor-like organ in the closely related *P. lucifera*. Aerial, arboreal pupation behaviours, such as tree positioning to influence microhabitats (Lloyd 1997), pupal development rates and protandry (Gentry 2003) are provided for the semi-tropical FL *P. borealis* (877 km south of TN). Cooler Massachusetts climate (1324 kilometers north of TN) *E. corrusca* adult wintering and mating behaviours and survival rates are described by Rooney and Lewis (2000). Like many temperate region fireflies, *E. corrusca* and *P. borealis* have well-defined seasons of appearance and peak activity (Faust 2004, Faust & Weston 2009). In East TN, adults of *E. corrusca* appear in Jan-Feb and are rarely seen after May; the last instar larvae of *P. borealis* also appear in February with few adults seen after April. Local species representatives can be found in the Great Smoky Mountains National Park collection (GSMNP), Gatlinburg, TN (Mayor 2006).

Active in the daylight and often found below 3 m on tree trunks, the adults of *E. corrusca* and larvae of *P. borealis* are easily accessible for live studies in their natural habitat before the other common fourteen-plus East TN species (Faust & Weston 2009) emerge. Both firefly species appear to return to the same or nearby trees each season. These fireflies, widespread but less abundant than other local firefly species, are found in loose colonies on the margins or within the secondary, mixed hardwood forests of East TN, which are predominantly Appalachian oak-hickory type (Fig. 3).

This longitudinal study, spanning four years of focused studies and 19 years of field observations, documents the continuity and stability of these populations and presents emerging threats, particularly to the Knox County population which was the primary study site.

The study site habitats shared by both species are described together. The species details are dealt with separately. Regional differences and similarities are discussed and new information presented on selection behaviour (of host trees and directional orientation), aspects of development (life event dates, mortality factors and survival rates) and reproduction (sex ratio, larval/pupal seeking and tending, mate guarding, flashing and mating). Average calendar dates and modified growing degree day values (mGDDC) for life stage landmarks (Faust & Weston 2009) are offered when significant as a reference point for future comparisons to far distant populations. This paper was presented at the August 2010 Malaysian International Firefly Symposium. New data from the 2011–2012 season has been added to augment the results reported at that meeting. It is hoped this information will further the understanding of these adaptable early season species.

METHODS

Study sites

There were five study sites in East TN, USA, i.e. Knox (35°56'38.13"N 84°12'15.97"W,

274-305 m a.s.l.), Blount (35°36'55.05"N 83°40'07.32"W, 640 m, GSMNP), Sevier (35°39'02.79"N 83°34'32.33"W, 731 m, GSMNP), Jefferson-Grainger (36°06'23.64"N 83°37'58.11"W, 305 m) and Morgan (36°12'11.52"N 84°35'17.14"W, 407 m). Fireflies of both species have been found at these sites every year since I started observing them in 1992. Knox County, the primary site, was 46–66 km from the other four sites. The Knox site consisted of a 4.3 km circular loop covering 15 ha of maturing species-diverse (i.e., more than 111 species) Appalachian oak-hickory type secondary forest (Oswalt 2007) that borders privately and publicly owned lands (Fig. 3). The four secondary sites are generally similar in habitat and forest composition, but the two GSMNP sites are higher in elevation (Faust 2010) and the Jefferson-Grainger site includes an alluvial floodplain forest. Using five sites enabled comparing the fireflies' tree usage and emergence patterns in slightly varying habitats within the same region. Within the sites, *Ellychnia corrusca* colonies have been found at the same locations every year and while there may be slight shifts in trees used, the number of trees occupied is fairly consistent. At Knox, the main colony occupies 29 trees located on a 0.5 ha karst hilltop rim while a smaller colony is found on 2 trees 0.5 km away. The circular 0.37 ha Blount site has a large colony occupying 20 trees and a smaller colony on 6 trees 1 km away.

Field sampling

The studies for the two firefly species took place over four seasons (2008–11). Outside the firefly season from May–January, monthly site visits recorded sightings of adults or larvae on trees or the ground. The sites were visited weekly from the first appear-

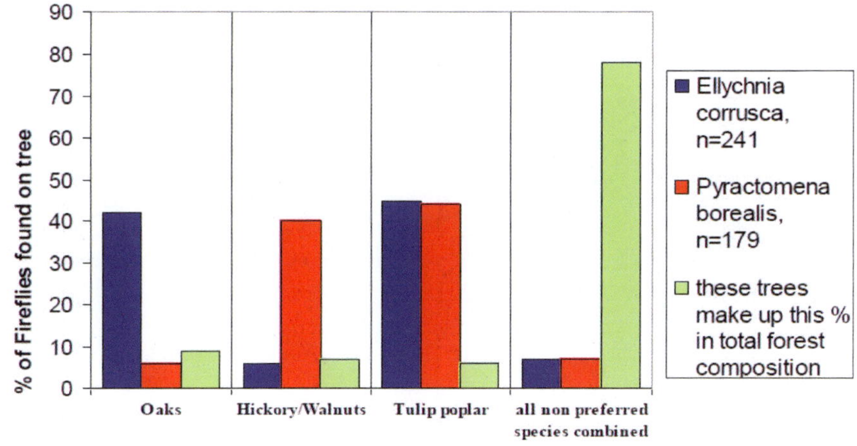

Figure 3 Preferred trees (9 species) utilised by *Ellychnia corrusca* and *Pyractomena borealis* in relation to the composition of a typical Appalachian oak-hickory forest with ≥111 species.

ance of *E. corrusca*, in Jan or Feb through early May (last sightings for *P. borealis*). During the first few surveys for each firefly species, the season's colony trees were identified and marked. Colony trees of previous years were checked first, followed by neighbouring trees. In the first year of this study, *P. borealis* exuviae persisting on trees used in 2007 identified previous colony trees for that firefly species. Searches for additional colony trees within the sites were carried out each season. For all colony trees, the following were recorded: species, diameter at breast height (dbh), height, vertical position in canopy layer (canopy or understory), bark characteristics (smooth, scaly, rough or furrowed; if furrowed, depth of furrow measured), and location in relation to amount of solar radiation received (sunny or shaded).

Surveys at secondary sites were not continued if fewer than five individuals were found during the first weeks of emergence for that particular species. In addition to the weekly site visits, the primary site (Knox) was visited 1–3 times daily from the week of first appearance. For both firefly species, site visits recorded the number of individuals found on each of their respective colony trees. For each individual, the following was also recorded: height from the ground; orientation in relation to:

1) aspect on tree (i.e., amount of solar radiation received, where the south-facing side is warmest),
2) the vertical line of the tree trunk,
3) other individuals; activity level (slow or fast-moving); sex (of adults for *E. corrusca,* when possible and at the pupal/adult stage for *P. borealis*). Aspects of behaviour during primary life events were recorded (quiescence, escape strategies, dispersal, mating and oviposition for *E. corrusca*; escape strategies, pupation, pupal-guarding, eclosion, courtship and mating for *P. borealis*). Mortality and likely causes were recorded, and also the presence of other insects resembling these fireflies.

As *P. borealis* courtship displays commenced at Knox, nightly walks along the loop trail were made to record the relative number of individuals in the treetops. Individuals could be reliably counted by their flashes because they usually flashed more than 5 m apart from their nearest closest neighbours. Towards the end of the courtship period as the trees leafed out (thus obstructing canopy view from beneath), flash counts were made from the creek bordering half the loop.

Additionally, in 2011 at Knox and Blount, all *E. corrusca* sighted were marked on their elytrae with distinct individual patterns of colour/dots/dashes with non-toxic ink (Uchida DecoColour™) identifying them by colony tree on the date they were first sighted. Daily (visual) recaptures were carried out to determine their movement (if any) to other trees. Live elytral length was measured using a Mayo™ caliper and ruler. Also in 2011 at Knox, about half of all *P. borealis* larvae sighted and later, all eclosing adults, were likewise marked and recaptured (visually) daily. Over the 4 years, random live measurements of larvae, pupae and adults were made to determine general live size ranges.

Accumulated modified growing degree-day (mGDD) values, also known as 30/10 C (or 86/50F corn growing degree days in US) and daily temperature data, were obtained from National Weather Service (NWS) station #406742 for all study sites except the higher elevation Sevier and Blount sites which used NWS #316341. For more complete explanation of this heuristic tool as it applies to fireflies and data tables see Faust and Weston (2009) and Fridley (2009), and the regional and federal climate centre microclimates websites (NRCC 2008-2010; NOAA 2008–10). Air temperatures were measured frequently and randomly comparing the microclimates of the south- and north-facing sides of colony and preferred trees.

Laboratory and controlled field studies

In 2010, *E. corrusca* adults (five males and five females) and a larva (a rare find) were observed in captivity for a week in April and September–October, respectively. The adults were placed in a mesh-capped two-litre glass jar lined with native soil and a few twigs positioned vertically. The larva was found glowing in the soil below leaf litter by a gravel road at Sevier at 9 pm, 23rd September.

In 2008 and 2009, separate mesh enclosures were put around four *P. borealis* female pupae where they were attached to their colony trees in order to determine when they first spontaneously flashed at males or artificial flashes after eclosion. Two captive virgin females were observed in 2008 and 2009 for copulatory behaviour and times. In 2011, 8 female and 5 male pupae were collected within several days of predicted eclosion and placed in natural position, pygypodia taped to sticks, in 1 quart mesh covered containers in a covered outdoor setting, within their habitat. Once eclosed, these 13 virgin adults were observed for 14 days for copulation events and times and remating behaviours and times. Two additional wild caught males were added to the remating studies the final week. Apple slices were provided as food. After 2-3 matings, females were returned to their pupal trees in order to preserve the local population. Males were either released at night or kept until death to determine typical male lifespan. Representative specimens of *E. corrusca* and *P. borealis* from this study are deposited in the Florida State Collection of Arthropods, Gainesville, FL. Photographs were taken with a Kodak Z740 Easyshare or an Olympus Stylus 720 and 770SW.

RESULTS

Life events of *Ellychnia corrusca*

Morphology

Like many diurnal fireflies with no lanterns, *E. corrusca* lacked paler elytral margins, yet retain the bright red and pale yellow markings on their wide (3-7 mm) pronotums (Fig. 2). The ridged dorsal elytral surface was densely covered by yellowish, olive green hairs. In the field, both sexes looked similar. Size and pronotum colour variation was common. The body and elytral length of specimens from Knox (16.2 mm ± 1.3 and 11.9 ± 1.1 mm, respectively; n = 19) were greater than the higher elevation (731 m, 640 m) Sevier and Blount adults (12.9 ± 1.7 mm and 9.4 ± 1.4 mm, respectively; *n* = 44). Females from 9 - 17 mm in size were seen coupled to males 9 mm in size (Fig. 4). In 2010, three of 164 newly emergent adults had bent or damaged elytra.

Figure 4 *Ellychnia corrusca* complex has large size variation. Coupling occurs on trees by day.

Larvae

Robust, dark bodied *Ellychnia corrusca* larvae, rarely seen, were covered in bristly coarse setae. One glowing larva, found in September, white and pink from recent molting, had darkened by the next morning (Fig. 5). It measured 15–17 mm long and 3.5 mm wide and readily ate small chopped live worms 6 times in 21 days. It used its pygypodia (caudal tail organ) for grooming and to pry open a fresh water snail operculum. It did not attack a more fragile *Photinus* sp. larva kept in the same enclosure. They often fed together, though the *Photinus* larva deferred to it. After six weeks, it died from unknown causes.

Abrupt appearance and vertical emergence of adults

From late October to early February, few to no adults or larvae were evident either on the ground or tree trunks. Adults were first sighted at one or more of the sites on 14, 10 and 18 February in 2008, 2009 and 2010, respectively, and on 20 January in 2011. At Knox (29 trees) in 2010 (Fig. 6), no adults were seen until 20 February when 44 were found. The next day's total was 104. Smaller groups emerged over the final week in February, usually on clear sunny days, with a late group of seven appearing on 29 March bringing to a total of 217 adults for that site and 299 for all sites.

Figure 5 *Ellychnia corrusca* late instar larva feeding on a fresh earthworm section.

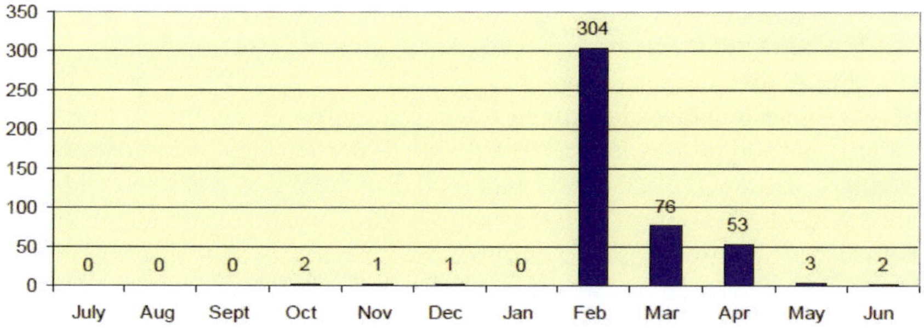

Figure 6 Visible presence of *Ellychnia corrusca* adults monthly for one year at Knox Co. colony, in flight from July 2009 - June 2010 and on trees from July 2009 - April 2010.

When emerging at temperatures below 5°C, often with snow-covered ground, the adults moved slowly. Mean height up the tree trunks increased daily from 66 cm (on 18 February) to 88 cm, 152 cm, and 170 cm on Feb 21 (range 15-152 cm, 22-243 cm, 96-307 cm, 0-320 cm respectively) at the Morgan and Knox sites with temperatures at appearance ranging from 0-16°C.

They were oriented facing a southerly direction (mean 182 ± 21°, n = 299) on the trees, and on each tree were vertically oriented to one another (Fig. 7). In 2010, at Knox, over 95% of 217 adults followed this vertical orientation to one another. The rest appeared scattered but still climbing the southerly aspect of the trees. On subsequent days, the vertical arrangement disappeared as the majority of adults scattered and climbed out of sight up the trees. In 2011, at Knox and Blount, 90% of the final tally of adults (n = 123) emerged more gradually from 29 January – 1 March with peak numbers reached by mid-March. Orientation was 189 ± 19°.

The 2012 season had an unusually mild winter and early spring. First *E. corrusca* emergence of 30 climbing adults was observed on 29 December 2011 at the Blount site, with subsequent dates of dispersal and coupling beginning 2-4 weeks earlier than previously recorded (by 19 March).

Figure 7 *Ellychnia corrusca,* six adults circled in black, abrupt appearance in February with group vertical tracking up the colony tree.

Colony trees

From 2008-10, oaks (*Quercus velutina, Q. alba* and *Q. coccinea*) and tulip poplars (*Tulipifera*

Larvae/ occassional adult	Few or no *Ellychnia* evident	Emerge & climb tree	Quiescent on tree trunk	Flight / dispersal	Coupling / ovipositing	Adult / death
May - November	June - January	February - early March	February - mid March	Mid - late March	April	Late April - early May[1]

[1] Early June for high elevation.

Table 1 *Ellychnia corrusca* general life cycle calendar, East Tennessee.

liriodendron) were colonised by 87% (n = 241) of adults. These tree species comprise only 16% of the total forest composition of trees >2.5cm dbh (Oswalt 2007). At Knox, in 2010 and 2011, 83% (n= 29) and 80% (n = 10) of the colony trees, respectively, were oaks while at Blount, 84% (n = 26) of trees colonised were tulip poplars, which are abundant there. Colony trees, 60 to 80 years old, were mature, living and dominant canopy trees found in open deciduous forests. They shared direct exposure to winter sun for most of the day, had rough, furrowed or scaly bark (furrow depth 12 ± 3.6 cm, n = 27), were canopy (25m+) trees and relatively large dbh (57 ± 16 cm) compared to the many other tree species in the surrounding forest. From 2008-2011, lightning, high winds and disease killed 4 of the colony trees.

Quiescence

The quiescent phase was from first appearance to mid-March (Table 1). The adults often rested head-down in south facing bark furrows, with no apparent vertical orientation to one another. They were minimally active (crawling) on any sunny above-freezing day. Marked adults (n = 123) in the recapture study frequently moved up and down the trees, as much as 3 meters in 30 minutes especially on sunny days. Up to 24 days after the first sighting, 96% of the 54 recaptured adults were on their original tree. The remainder were on nearby trees (< 3 m away), having dropped to the ground and disappeared when initially marked (which threatens them). In the final 2-3 weeks of quiescence before dispersal, no recaptures were recorded at Knox and fewer than twelve were sighted at Blount. They may have climbed up into the colony tree canopies. Few were seen in late quiescence until flight, dispersal and coupling began. In 2012, adults were visually recaptured as long as 66 days past initial marking.

Dispersal, mating and ovipositing

Dispersal, recognised by the reappearance of fireflies in flight or active on both colony and non-colony trees, occurred in mid-March at air temperatures above 15°C. First observed flight in 2010 and 2011 was on 8 March at 20°C and 17 March at 23°C, respectively. On very sunny days where south-facing trunk temperatures exceeded 25°C, adults were often seen on the north sides of the still-leafless trees where temperatures were 5-10°C cooler. From 2008-11, coupling started in April within 119-222 mGDDC (215-400 mGDDF), regardless of warm weather before that time. At Knox and Sevier, the most active coupling period in 2010 and 2011 was 9-10 April (10 couples and 11 singles recorded) and 29 March – 9 April (11 couples and 49 new unmarked singles recorded), respectively. In 2011, 23 and 19 colony and non-colony trees, respectively, were used for courtship and coupling. Mating was not observed on the ground. Flying individuals were most often seen after 12 noon when it was warmest. By April's end, few remained and no coupling was observed at Knox while at Sevier, small numbers (<10) persisted through the first week of June in 2009, 2010, and 2011. From June-Dec, adults were rarely seen (Table 1 and Fig. 6).

In the laboratory, multiple matings between the 5 males and 5 females were observed for both sexes, in Stage 1 and 2 positions (see Lewis & Wang 1991). Coupling (9 observed in 7 days) took place vertically on twigs when available (otherwise on the soil), commencing before sunrise and lasting 2 to more than 6 h. Females laid off-white eggs measuring 0.5-0.9 mm (n= 27) singly in the soil on successive days. Under magnification, the egg surface appeared slightly rough and sticky.

Escape strategies

When observed too closely or feeling threatened, their two primary escape strategies were actively crawling up the trunk or dropping to the ground on their backs, becoming well-camouflaged in the leaf litter. Escape climbing took place even on cold (<0°C) days, albeit slower than on warmer days (especially >20°C). When marked, they exhibited distress by dropping to the ground, opening and closing their elytra or attempting to rub their bodies on the bark. When handled or when crawling, they often deposited a clear droplet of liquid from their terminal segment. They exuded white beads of reflexive bleeding from the elytral margins when severely threatened. Flying, observed only after dispersal, was seldom used as an escape strategy.

Figure 8 Left to right: A white phorid (*Apocephalus antennatus*) larva emerges as its female *Ellychnia corrusca* host dies; phorid pupae; adult phorid. Phorid flies are parasitoids of *E. corrusca* and *Pyractomena borealis*.

Figure 9 Mimicry complex on/near trees colonised by *Ellychnia corrusca* and *Pyractomena borealis*: **1-5**. Cantharidae (8-12 mm); **6**. Lycidae (9 mm); **7**. Meloidae (6 mm); **8**. Lampyridae (*E. corrusca* on top, 15 mm; *P. borealis* beneath, 17 mm); **9**. Carabidae (15 mm); **10**. Pyrochoidae (12 mm); **11**. Silphidae (20 mm); **12**. Cerambycidae (9 mm); **13-14**. Arctidae (20 mm); **15-17**. Elateridae (10 mm); **18**. Pedilidae (7 mm).

Predation and mimics

Predation was recorded early (28 February) in the season for 2010 and 2011 and parasitism late (22 April) in the 2011 season. In 2010, two of four males taking shelter in a hole at the base of a *Quercus alba* tree were preyed on and in 2011, three unmarked elytra were found at the base of colony trees in thick leaf litter. Bite marks and/or tears on elytral remnants were consistent with rodent or bird predation. In 2011, 32 phorid larvae (Brown 1994) that emerged from a female that died shortly after, were later identified (Brown, pers. comm.) as *Apocephalus antennatus* Malloch (Diptera; Fig. 8). Beetles and moths from several families with mimic-like colouration and patterns were often found on or near the firefly colony trees (Fig. 9).

Life events of *Pyractomena borealis*

Fall larvae

Over the four years, a total of five late instar larvae were observed in September and October (outside the *P. borealis* season). They were found glowing intermittently on leaf litter and vegetation up to 20 cm from the ground.

Climbing-searching larvae

The first sightings of last instar, searching-climbing larvae occurred during the day on 1st March, 10th, 20th and 21st February in 2008-11, at temperatures of 12°C, 10°C, 16°C and 19°C, respectively. In the very mild winter of 2012, larvae climbed over a month earlier (Jan 22), first flashed and fully eclosed up to 3 weeks earlier (19-25 Mar) than previous years. Male larvae climbed-searched 15 ± 8 d (n = 68, 2008-11) earlier than the females (n = 40). The pink-tinged larvae measured 16-26 mm (live length, n = 20). They were narrow-headed, cryptically coloured, mottled light yellow to dark brown. They climbed up from the leaf litter and actively investigated the south-facing furrows low on the trunks of their chosen trees. They got no higher than 1 m up the trees in the earliest searching phase in February and often disappeared early in the season. On very cold days (< -8°C), they remained motionless in the furrows. They relocated every few days in search of more southerly or protected sites.

By early March they were head down with dorsum facing outward, and had begun to anchor their pygypodia within the furrows or under bark scales (Figs. 10 & 11). Four pupated 'backwards' with the dorsum facing the tree and/or (exposed) outside the furrow. Anchoring was often permanent within 3 days. Larvae disturbed in the first two days were found to abandon the site (12 of 21 larvae moved when marked in 2011) instead of going on to pupate. Once attached, if disturbed, they glowed and arched backwards and up.

The pupation sites were at a mean height of 147 ± 46 cm (*n* = 107) and oriented at 170 ± 23°. In 4 years, the lowest pupation site was 28 cm and the highest observed was 300 cm. Males pupated 12 ± 7 d (n = 68, 2008-11) earlier than females (n = 40) but their pupation heights appeared no different. Mean furrow depth for these pupation sites was 10 ± 3 mm (n = 53).

Relative temperature measurements taken randomly at the actual pupation sites throughout the firefly season averaged 16°C (*n* = 320) compared to the opposite, shadier side of the trees, which averaged 14°C. On sunny days before the trees leafed out, the temperature difference between the north- and south-facing pupation sites often exceeded 13°C.

During the larval climb and pupation period, temperatures ranged from -1°C to 31°C. The range for pronotal width was 4-7 mm and body length was 11-26 mm (n = 45), with the largest larvae and pupae (22-26 mm, n = 21) eclosing into the largest female adults (20-22 mm).

Pupation

By early April most of the larvae had pupated into the characteristic black pupae with larval exuviae still attached to the pupal dorsum (Fig. 10). Two pupae at pupation sites that reached extraordinarily high temperatures of 42°C (on 17 April 2010) failed to eclose. The two pupae that pupated backwards and outside a furrow did not eclose but the two that pupated in exposed locations did eclose.

Eclosion

Males began to eclose late in March, 9 ± 7 days (n = 68, 2008-11) earlier than the females (n = 40). The sexes could be differentiated by their lanterns (Fig. 12). On cold days with inclement weather (< 5°C), newly-eclosed adults stayed next to their pupal exuviae for up to 24 h as they hardened and darkened while on hot days (> 25°C), they quickly moved to a shadier location on the trunk.

Among the marked larvae in 2011, males and females began to eclose on 27 March and 3 April, respectively. By 8 April over 50% of the females (n = 15) had eclosed, with the last male and female eclosing 9 April and 14 April respectively, bringing to a total of 35 males and 15 females (Fig. 13). All were re-marked soon after eclosing. Eight males disappeared after marking, but the other males remained on their pupation trees. There was no flashing from these first-eclosing, day-active searching males for over a week. These males, diurnal at this stage, methodically searched the trunks for larval or pupal females by day and rested motionless in furrows at night (Figs. 10 & 11). Every pupa, furrow or scale 3 m up the trunk was investigated. Upon locating a female, the male clasped her with his legs, pressed his

Figure 10 Life stages of *Pyractomena borealis*. Protandrous males guard female larvae and pupae and attempt coupling as female ecloses.

pronotum to hers and became motionless. Pupae, when approached by a potential guard male, arched back and forward rapidly (12 times in 2 min in one instance) before being clasped and subdued by the male. A few male pupae (2 in 2011) were guarded, but only briefly (0.25-3.00 h). All pupae guarded for over a day eclosed as females (2008-11). Competing males piled on top, grappled with, and often displaced the guard male, which would then be forced to search for another female pupa. Males also used their pronotum to pry under or push away other males. This competition intensified as females approached eclosion, with frequent power reversals occurring among the increasing numbers of males. Female pupae 24 h from eclosing had 3-7 males competing for them (n = 9).

From 2008-11, 79.1% (n = 43) of larval and pupal females were guarded by one or more males from 1-14 days until eclosion (Figs. 10 & 11), which took about 30 min. From 2008-11, 74.6% (n = 150) of eclosing adults were males and this male-biased sex ratio was significant (p<0.0001, exact binomial test). Preliminary molecular studies on this local population found no evidence of *Wolbachia* bacteria (Werren et al. 1995) as influencing this ratio (F. Vencl, pers. comm. 2011) as has been found to occur in some Korean firefly species (Jeong et al. 2009).

Adult live mean body length, elytral length and pronotal width was 17 ± 2 mm, 11.5 ± 1.4 mm and 4.5 ± 1.2 mm, respectively (n = 58). Females, with distended abdomens protruding 2-4 mm beyond the elytrae, often measured longer than males (20.1 ± 0.9 mm, n = 9 vs. 15.8 ± 1.7 mm, n = 35). Elytral length was more similar: males 11 ± 1.2 mm; females 12 ± 1.6 mm (Fig. 12). In 2011, three of 88 newly eclosed adults had deformed or missing elytrae and/or wings. This amount of deformation was representative of previous years.

Figure 11 *Pyractomena borealis* males tend female larvae (A & B) and pupae up to two weeks before eclosion. Disturbed males may abandon the female, but return when threat is gone. Two and four males (C & D), some marked, grapple over a female pupa.

Figure 12 Lantern of male and female *Pyractomena borealis*. Two lanterns on the male and four smaller lanterns visible on the female.

Display and mating

Typically, a newly-eclosed (still white and soft) female (n = 18), would be mobbed by her guard and competing males. She would rapidly leave her pupation site, seeking cover under a bark scale or crevice, with 1-3 males riding her back. Copulation commenced before her elytra had hardened and darkened and no flash dialogue was involved (Figs. 10 & 11).

From 2008-2011 first flight and flashing started on 10th, 12th, 11th and 6th April respectively after half the females had eclosed and climbed up the tree disappearing in the canopy. In 2011, males then ceased patrolling the trunks and even stopped guarding remaining pupae, decreasing from 46 guard males to 2 by 12 April with one female pupa left. Males displayed high in the bare or newly leafed-out treetops, from 30 min after sunset till sunrise with peak display 45-90 min after sunset. Temperatures were 14-22°C during display times. Peak nights recorded 50-120 observed individuals along the 4.3 km trail, showing that this species, though common in dark forest habitats, is not abundant. Three flash patterns were observed. The most common was the male's single 0.3-0.5 s flash every 2-3 s. They flashed as they flew slowly around the canopy margins. Less common was the flash pattern similar to the first, but with the lantern dimly lit in between flashes. The

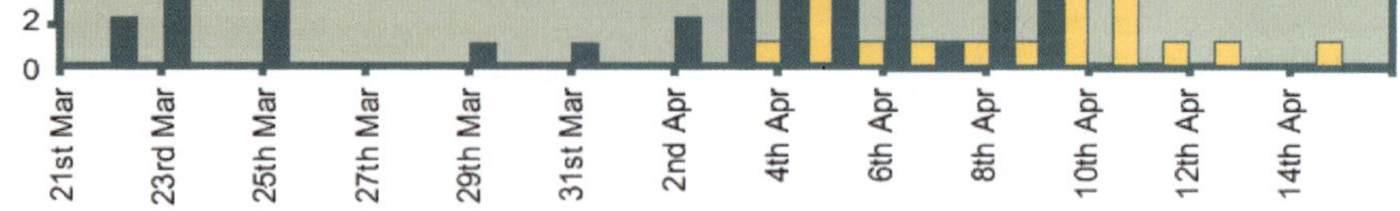

Figure 13 *Pyractomena borealis* male and female eclosion, pupal guarding and display dates for 2011. Note cool spell <10°C, with little activity, typical for this time of year, between 24 March and 2 April.

rare female seen flashing (n=7, in the wild and in captivity) responded immediately after the male flash with a single or bimodal flash lasting ca. 0.70-1 s, with the initial flash brighter.

In the controlled field study, the 4 newly eclosed virgin females did not respond to artificial or male flashes until the third night after eclosion. In the 2011 mating/remating studies, the 8 virgin females were not initially receptive to the advances of males introduced to the cage. Females fled, squeezed into crevices and tucked their tails under their bodies when pursued. Despite this avoidance behaviour by the females, males ultimately copulated successfully in 14 of the 16 observed matings in captivity, but less often in the wild where the female often escaped the pursuing male (unquantified pers. obs.) Copulations from 2008-2011, usually remaining in Stage 1 position (Lewis & Wang 1991), lasted an average of 67 ± 64 h (n=17) (range 3-214 h), with mean virgin pairings lasting 66 ± 62 h (n=11) and mean re-matings lasting 84 ± 74 h (n=5). Males stayed in copula even when probed or misted with water. Once copulation began, competing males left. If a new male arrived, brief skirmishes occurred as he tried to take the female, but did not succeed. Both males and females re-mated readily (n= 8). After copulating, the males glowed continuously. When released at the base of their colony trees, they continued glowing, but after 5 min switched to the characteristic flash pattern, all the while climbing up the trees. In the controlled field study, the first male to eclose guarded 2 females for 9 days, coupled for 65 h in captivity with a third and died on the 24th day. One of the last males to eclose also survived 24 days, guarding 3 females for 5 days and mating twice (160 h). Females, when returned to their pupal tree (placed on the ground near the tree), sought the tree and climbed upward. One female with an extraordinarily distended abdomen climbed to over 5 m, flew down to a nearby maple and disappeared in the new foliage.

Predation and parasitism

In 2011, three newly-anchored larvae were preyed on as evidenced by remnant anchored pygypod stumps. Possible predators were rodents, birds and spiders. From 2008-10, 14 of 94 pupae (15%) died. Twelve were eaten by spiders. A Johnson jumper, *Phidippus johnsoni* group, and the star bellied orb weaver, *Acanthepeira stellate,* were observed eating pupae and newly-eclosed adults, and leaving silk-wrapped body parts attached to the tree. Some partially-eaten pupae survived to eclose albeit with deformed elytrae and pronotums. From 2008-10, 17.4% (n = 132) of the larvae and pupae did not progress to adults. A pupa that fell naked to the ground after its exuvia was eaten by an isopod woodlouse, *Porcellio scaber,* successfully eclosed as a male 7 days later in the laboratory. Phorids, *Antennatus apocephalus* (Lloyd 1973, Fig. 8*)* parasitised adults from 2008-11. Predator *Photuris* fireflies were not a significant threat to these early season fireflies as local *Photuris* were not active until May. Mimics were abundant throughout the display season (Fig. 9).

First Larval climb*	First pupa	First male	First female	First flash*	Peak display	<10% Remain
Jan 22 - Mar 1st	Mar 4th - 25th	Mar 22nd -Apr 8th	Apr 3rd - 11th	Mar 19th - Apr 12th	Apr 14th - 22nd	Apr 24th - May 2nd
mGDDC 0-1.4	mGDDC 18-59	mGDDC 103-152	mGDDC 134-174	mGDDC 125-176	mGDDC 218-267	mGDDC 307-330
mGDDF 0-2.5	mGDDF 32-106	mGDDF 184-272	mGDDF 239-310	mGDDF 225-315	mGDDF 397-477	mGDDF 549-589

Table 2 *Pyractomena borealis* first-sighting life-stage landmark date and degree-day range for 2008, 2009, 2010, 2011 for Knox, Jefferson, Morgan, Sevier Co., TN. One day = 3 yr. mean of 6.5 mGDDC in April. * 2012 larval climb and first flash included. Average durations for total population provided in text.

Dates and degree-days

From 2008-11, over half the nights in April had cold (< 12°C) or inclement weather (e.g., Beaufort wind force 3+, rain, snow) unfavorable for flashing display. The larval climb appeared related to relatively warm sunny days in February. Degree-days, were useful for prediction of pupal and adult life stages and maximum display days (Faust & Weston 2009). From 2008-11, both sexes combined (n = 78) averaged eclosion on April 10 or after 16 ± 5 days of pupation or 86 ± 16 mGDDC from pupae to adult. The 18-year average for the occurrence of flash display behaviour at Knox was 19 March – 24 April, mGDDC 119-234. Weather is extremely variable in April e.g., 5 mGDDC/day in 2008 and 2009, and 7 mGDDC/day in 2010 and 2011, which is seen in the varying pupal durations in calendar days, but more constant durations in degree-days (Table 2). The higher elevation Sevier site had a later season with later calendar dates but similar degree-day ranges.

Pupation trees

Pupation trees had certain traits: scaly or furrowed bark (10 ± 3 mm, n = 53), tall (21-30 m), larger dbh trees (49 ± 14 cm, n = 73) with direct southern exposure to the sun (located at forest margins or open gaps). Tulip poplars, hickories (*Carya pallida, C. cordiformis, C. glabra* and *C. tomentosa*) and walnuts (*Juglans nigra*) were used by 84% (n = 179) of the larvae; several oaks (*Q. velutina, Q. alba, Q. phellos* and *Q. bicolour*) were used by 6%; three other tree species (pine: *Pinus virginiana*; persimmon: *Diospyros virginiana;* and winged elm: *Ulmus alata*) were used by 2.5%. Two 60 year-old Eastern Red Cedars (*Juniperus virginiana*) with an abundance of stringy bark on smooth trunks, were the first record of that tree species as a pupation tree for *P. borealis*. These cedars held eight pupae in 2011. Of the 20 trees at Knox used in 2008, 18 were reused 2-4 seasons. In 2011, 4 of 18 trees were new. Since 2008, 62% of preferred marked trees (n = 74) had one pupa, 37% had 2-7 pupae and 1% had 8-37 pupae. A lone sand hickory *C. pallida* surrounded by choking kudzu *Pueraria lobata* sheltered 37 pupae in 2011, the most observed pupae per one tree.

DISCUSSION

Trees used

The occurrence of both firefly species on particular tree species is likely due to preferential selection as these tree species are not the most abundant and comprise less than 16% of the tree species in these forests (Oswalt 2007) (Fig. 3).

In congregating Asian firefly species, GPS data on specific trees known to attract mating congregations is being documented by the Malaysian Nature Society (Mubah 2010), the Forest Research Institute Malaysia (FRIM) and others for the rivers of Sarawak and Malaysia (Nada et al. 2009; Wong 2009; Jusoh et al. 2010; Ohba & Wong 2012) and reflects the growing understanding of the critical importance of preserving larval habitat and preferred trees, and educating the public of this precious resource in both the East and West.

For the TN fireflies, which occupied rough-barked, large diameter trees whose trunks received plenty of sunlight, tree selection may be based on availability of trees that provide the most favorable conditions for the fireflies. Similar findings and conclusions were reported by Rooney and Lewis (2000) for *E. corrusca* adults and Gentry (2003) for *P. borealis* larvae in FL; (in contrast, see Lloyd's 1997 report of preference for smooth-barked trees). The mass of the larger diameter trees combined with south-facing, exposed trunks of the trees provide warmer spots needed for *P. borealis* larvae early in the season when it is very cold, while deep bark furrows provide shelter from rain, snow and wind and other weather extremes.

Calendar days and degree days

Because both species range over 2200 km north to south across the eastern US and parts of Canada, use of degree-day ranges combined with traditional calendar date ranges could be a more accurate method of determining local seasonal appearance regardless of population location.

As evidenced each year including during the record breaking warmth in 2012, firefly life stage dates in calendar days can vary by 3-4 weeks (Faust & Weston 2009), but the degree day ranges remain more constant, even in the record breaking warmth of 2012. *Ellychnia corrusca* appeared less influenced by accumulated degree days, though their coupling season was not reached until at ca. 100 mGDDC accumulated, despite periods of warmth earlier in the season.

For *P. borealis*, both average calendar dates and mGDDs were useful in predicting when certain life stages (pupal duration and eclosion by sex and peak display/courtship) would occur and when the season would end (Table 2). Increasing survey frequency during the predicted times for those life events helped capture important information that may otherwise have been overlooked. Using degree-days as a prediction tool could guide well-timed collecting efforts when limited by National Park regulations and prevent over-collecting small local populations.

Predation and parasitism

Adults of *E. corrusca* suffered early winter losses to probable rodent activity as evidenced by the elytral remnants with bite marks found at the base of the colony tree. In the milder winters of East TN, rodents such as mice and shrews are active year-round and could pose a threat especially to *E. corrusca* near the ground. Rooney and Lewis (2000) also found elytrae at the base of colony trees and suspected rodent activity. The abrupt vertical emergence of the TN *E. corrusca* could help minimise losses to rodent predation by allowing the fireflies to quickly ascend *en masse* up the trees and away from the leaf litter. For *P. borealis*, the higher pupal losses to spider predation and no (observed) predation of larvae raises the question as to why this disparity exists. Larvae have chemical defenses (Lewis & Cratsley 2008) but it is not known whether the same is true for the pupae. Early emergence in TN appears to be an effective strategy by both firefly species to avoid predation by *Photuris* fireflies, which become active later and are important predators of *P. borealis* adults elsewhere (Buschman 1974). Phorid parasitism of both species appeared to increase as the season progressed (Fig. 8).

Additional discussion for *Ellychnia corrusca*

These fireflies appear to overwinter as either adults or pupae somewhere other than on exposed tree trunks and make a sudden initial appearance at the base of the tree (and then climb up) during the first relatively warm days of Jan-Feb, despite more below-freezing weather being a certainty at this point of winter. In contrast, the Massachusetts (MA) *E. corrusca* adults much further north, overwinter on the trunks of their black oak colony trees (Rooney & Lewis 2000) beginning in early fall. The more northerly Pennsylvania (Butler Co.) *E. corrusca* (pers. obs.) appears to share the overwintering habit of the Massachusetts firefly. In contrast, Hackett et al. (1996) mention collecting *E. corrusca* in late winter 1983 "while they were emerging from their overwintering sites at the base of a white oak tree in Beltsville, Maryland." The more southern Maryland *E. corrusca's* sudden appearance at winter's end more closely parallels the TN *E. corrusca* habit. Further studies are needed to determine whether these regional differences in emergence and overwintering behaviours suggest distinct species, univoltine vs. semivoltine life cycles perhaps influenced by the milder climatic conditions in TN, or adaptive behaviours governed by winter conditions and predation risks for different locales. Since *E. corrusca* are a species complex (Fender 1970) instead of a distinct species, determining differences in larval development, adult size variations, overwintering habits and adult emergence times could help in species discrimination. No adults have yet been observed feeding in TN at sap flows or on flowers as they have in the more northern range. Perhaps the arboreal overwintering habit of the MA adults requires more energy. Captive TN *E. corrusca* will however, feed from an apple slice, sometimes choosing the apple over mating.

The clear drops of liquid that both sexes deposit from their terminal segment when disturbed and/or climbing are in addition to the white beads of reflexive bleeding previously noted and described (Williams 1917; Rooney & Lewis 2000, respectively). This clear liquid may be excreta, or a chemical cue promoting sexual or social communication. The vertical upward migration on the days of initial emergence (Fig. 7) suggests these fireflies are following one another up a particular route of each tree and that something more than just the direction and heat from the sun influences this route

Further studies could focus on what actually occurs during the quiescent 4-6 wk after initial appearance but before flight and coupling. What is observed is that they search the trunk or rest head down in the furrows and in the later weeks of quiescence, disappear into the canopy. The MA *E. corrusca* are known to be sexually mature early in the season (Rooney & Lewis 2000) so perhaps unseen pre-flight coupling is occurring in the canopy. Pre-flight mating, if true, would also reduce risks associated with flight, e.g., exposure to the elements and predators, for these weak flyers.

Additional discussion for *Pyractomena borealis*

The 'fickle' behaviour of larvae in their constant search for a better pupation site has also been observed previously (Lloyd 1997). TN larvae were especially mobile when first climbing in January and February and often changed locations but not trees before settling on a permanent pupation site usually by late February or early March. Some larvae may disappear above the surveyed 3 m mark, where others succumb to predation or the elements. Larvae pupated < 2 m high on the only two smooth but stringy barked trees, yet the scaliest, roughest trees that are harder to climb often had pupation sites as high as 3m, a counter-intuitive finding. Further study is needed to determine if TN larvae choose only furrowed trees for pupation, or pupate out of sight above 3 m on rough and smooth-barked trees. Red and sugar maples (*Acer rubrum* and *Acer saccharum)*, which are smooth-barked species until extreme age, are the most common trees in the local forests (Oswalt 2007) comprising 17% of all stems >2.5cm dbh, yet *P. borealis* have not been found on any maples (Fig. 3). Instead, 90% of pupae are found on poplars, oaks, hickories and walnuts whose combined species comprise only 21% of the forest (Fig. 3).

Larvae avoid most of the forest floor vegetation that shades the lower tree trunks by climbing up the trunk 1-2 m into the more direct warming rays of the sun. Climbing higher has also been proposed as means to stay above high water levels in flood-prone, low-lying forests in FL where *P. borealis* is found (Lloyd 1997). This may also apply to the *P. borealis* that colonise trees in the alluvial forests in TN, though many of the preferred trees are located on high ground.

That the male larvae climb, attach, pupate and eclose earlier in the wild on average than females concurs with findings from a previous laboratory study (Gentry 2003) where pupal durations were determined at constant temperatures. Findings of Al-Saffar et al. (1996) and Weston and Diaz (2005) indicate that varying temperatures can actually enhance maturation in some insect species. Could it be possible that the males obtain further developmental advantage by climbing and pupating earlier in the season when the temperatures are more extreme and variable? Is it possible that the guarding males provide some measure of thermal protection to their tended female pupae thus influencing female development (Figs. 10 & 11)?

Protandry gives the males much time to locate and guard potential mates, which are often coupled immediately after the female's eclosion. This diurnal mating strategy, taking place during the relative warmth of the day, bypasses flash communication and may be an important adaptation to the nighttime conditions that are often unfavorable to courtship displays. Locating females without needing to fly or flash may increase males' mating success and reserve their energy for photic display on the few warm nights available when flight conditions are ideal. This dual strategy approach ensures the males will have mating opportunities regardless of the unpredictable late winter/early spring weather. While eclosing virgin females had little control over mate choice, in rematings, depending on the situation and the male, females displayed both acquiescent and effective avoidance behaviours. The prolonged (up to 8 days) copulatory mate-guarding seen in *P. borealis* is common in insects and is a male adaptation to avoid sperm competition (Tompkins & Simmons 2000).

Future

In the past 19 years, the Knox fireflies have faced many challenges affecting their once dark, forested rural habitat. The clearing of forested areas that the fireflies inhabit and in which they display is a major threat, with 15% of the primary Knox site lost in the past four years alone. While daytime mating incidentally precludes *E. corrusca* from the negative effects of artificial lighting, the nocturnal courtship phase of *P. borealis* is not spared. Homes that have been built in these areas have introduced light pollution and pollution from landscaping agrochemicals that may be contributing to the apparent decline in firefly populations. New shopping centers, a football stadium and schools built less than 2 km away now illuminate the night skies to the point that a flashlight is no longer necessary on some nights while walking the firefly survey trails. Habitat loss and light pollution are also serious threats to Asian firefly populations (Wong 2009; Nada et al. 2009) and fireflies worldwide. Other imminent threats to the TN fireflies concern threats to their colony trees. In addition to natural mortality each year due to high winds, disease and lightning, the invasive kudzu vine *Pueraria lobata* threatens to smother many of the preferred and colony trees. The Thousand Cankers disease *(Geosmitha* spp.) that is spread by the walnut twig beetle, *Pityophthorus juglandis* (Coleoptera: Scolytidae) and is deadly to black walnuts, has recently been found infesting walnuts in TN (M.T. Windham, pers. comm., August 2010).

Future studies of mechanisms governing the abrupt appearance and vertical mass tracking up colony trees might provide new understanding of chemical cues, early season adaptations and regional differences in behaviour of the *E. corrusca* complex. The described observations of male *P. borealis*' daytime search - and guard-behaviour of larva/pupal females, their copulation without courtship dialogue early in the mating season, and their nocturnal habits and flash displays in the latter half of the season, provide a first insight into a dual mating strategy of a lampyrid. With their preference for trees of a certain size, furrow depth, species and specific location, both of these adaptable, hardy firefly species nonetheless remain vulnerable to natural and man-made changes to their environments. It is hoped these studies will further the appreciation of the importance of individual trees and habitat to the survival and health of larval, pupal and adult firefly populations and increase the understanding of these two species' remarkable adaptations to the habitat and climatic extremes of the early season.

ACKNOWLEDGEMENTS

My thanks to the hosts and sponsors of the 2010 International Firefly Symposium held in Selangor, Malaysia for providing such a wonderful opportunity for cooperation, collaboration, learning and sharing of information on the world of fireflies. Special appreciation goes to the Forest Research Institute of Malaysia and the Malaysian Nature Society. Thanks to Edgar Faust and Richard Evans for tree details, to Ted Faust and Paul Weston for technical assistance, Haden Wesley and Martie Wood for help in the field and The Great Smoky Mountains National Park (permit #FaustGRSM2009-SCI-0026) for access to two study sites.

REFERENCES

Al-Saffar, Z., Grainger, J. & Aldrich, J. (1996) Temperature and humidity affecting development, survival and weight loss of the pupal stage of *Drosophila melanogaster* and the influence of alternating temperature on the larvae. *Journal of Thermal Biology* **21** 389-396.

Archangelsky, M. & Branham, M. (1998) Description of the preimaginal stages of *Pyractomena borealis* (Randall, 1838), and notes on its biology (Coleoptera: Lampyridae). *Proceedings of the Washington Entomological Society* **100** 421-430.

Bousquet, Y. (1991) *Checklist of Beetles of Canada and Alaska.* Ottawa: Canada Communication Group. 430 pp.

Brown, B.V. (1994) Life history parameters and new host records of phorid (Diptera: Phoridae) parasitoids of fireflies. *Coleopterists Bulletin* **48** 145-147.

Buschman, L. (1974) Flash behaviour of a Nova Scotian firefly *Photuris fairchildi* (Barber) during courtship and aggressive mimicry. *Coleopterists Bulletin* **28** 27-31.

Buschman, L. (1984) Biology of the firefly *Pyractomena lucifera* (Coleoptera: Lampyridae). *Florida Entomologist* **67** 529-542.

Faust, L. (2004) Fireflies as a catalyst for science education. *Integrative and Comparative Biology* **44** 264-265.

Faust, L. & Weston, P. (2009) Degree-day prediction of adult emergence of the firefly *Photinus carolinus. Environmental Entomology* **38** 1505-1512.

Faust, L.F. (2010) Natural history and flash repertoire of the synchronous firefly *Photinus carolinus (*Coleoptera: Lampyridae) in the Great Smoky Mountains National Park. *Florida Entomologist* **93** 208-217.

Fender, K.M. (1970) *Ellychnia* of western North America. *Northwest Science* **44** 31-43.

Fridley, J.D. (2009) Downscaling climate over complex terrain: high fine-scale spatial variation of near-ground temperatures in a montane forested landscape (Great Smoky Mountains, USA). *Journal of Applied Meteorology and Climatology* **48** 1033-1049.

Gentry, E. (2003) On sexual selection in Florida's *Pyractomena borealis* (Coleoptera: Lampyridae). *Florida Entomologist* **86** 114-123.

Green, J. (1957) Revision of the Nearctic species of *Pyractomena* (Coloeoptera: Lampyridae). *Wasmann Journal of Biology* **15** 237-284.

Hackett, K.J., Whitcomb, R.F., French, F.E., Tully, J.G., Gasparich, G.E., Rose, D.L., Carle, P., Bové, J.M., Henegar, R.B., Clark, T.B., Konai, M., Clark, E.A. & Williamson, D.L. (1996) *Spiroplasma corruscae* sp. nov., from a firefly beetle (Coleoptera: Lampyridae) and tabanid flies (Diptera: Tabanidae). *International Journal of Systematic Bacteriology* **46** 947-950.

Jeong, G., Kang, T., Park, H., Choi, J., Hwang, S., Kim, W., Choi, Y., Lee, K., Park, I., Sim, H. & Kim, J. (2009) *Wolbachia* infection in the Korean endemic firefly, *Luciola unmunsana* (Coleoptera: Lampyridae). *Journal of Asia-Pacific Entomology* **12** 33-36.

Jusoh, W.F.A.W, Hashim, N.R. & Ibrahim, Z.Z. (2010) Firefly distribution and abundance on mangrove vegetation assemblages in Sepetang estuary, Peninsular Malaysia. *Wetlands Ecology and Management* **18** 367-373.

Lewis, S. & Wang, O. (1991) Reproductive ecology of two species of *Photinus* fireflies (Coleoptera: Lampyridae). *Psyche* **98** 293-307.

Lewis, S.M. & Cratsley, C.K. (2008) Flash signal evolution, mate choice and predation in fireflies. *Annual Review of Entomology* **53** 293-321.

Lloyd, J.E. (1973) Firefly parasites and predators. *Coleopterists Bulletin* **27** 91-106.

Lloyd, J.E. (1997) On research and entomological education, and a different light in the lives of fireflies (Coleoptera: Lampyridae: *Pyractomena*). *Florida Entomologist* **80** 120-131.

Mayor, A. (2006) *National Park Service. Great Smoky Mountains National Park.*

Gatlinburg, Tennessee. Discover Life in America (DLIA). All Taxa Biodiversity Inventory. http://www.dlia.org/atbi/species/Animalia/Arthropoda/Insecta/Coleoptera/Elateroidea/Lampyridae.

Mubah, M. (2010) Firefly night time assessment in Niah River. *Sarawak Fireflies.* http://sarawakfireflies.blogspot.com/2010/01/firefly-night-time-assessment-in-niah.html.

Nada, B., Kirton, L., Norma-Rashid, Y. & Khoo, V. (2009) Conservation efforts for the synchronous fireflies of the Selangor River in Malaysia. In *Diversity and Conservation of Fireflies. Proceedings of the International Symposium on Diversity and Conservation of Fireflies, Queen Sirikit Botanic Garden, Chiang Mai, Thailand, 26-30 August 2008* (Ed. by Napompeth, B.), pp. 160-171. Chiang Mai: Queen Sirikit Botanic Garden.

NOAA (National Oceanic and Atmospheric Agency). (2008-2010) http://www.cpc.noaa.gov/products/analysis_monitoring/cdus/degree_days/grodgree.txt.

NRCC (Northeast Regional Climate Center). (2008-2010) http://climod.nrcc.cornell.edu/climod/dd/.

Obha, N. & Wong, C.H. (2012) Observations on the external morphology and behaviour of *Pteroptyx tener* and *Pteropytx bearni* in Malaysia. In: *Firefly Conservation: From Science to Practice. Proceedings of the Second International Firefly Symposium, Selangor, Malaysia, 2-5 August 2010* (Ed. by Kirton, L.G., Lim, G.T. & Day, J.C.). *Lampyrid* **2** 80-86.

Oswalt, C. (2007) *Forest Composition of Oak-hickory Forest Type.* USDA Forest Service, Southern Research Station, Forest Inventory and Analysis Program.

Rooney, J. & Lewis, S. (2000) Notes on the life history and mating behaviour of *Ellychnia corrusca* (Coleoptera: Lampyridae). *Florida Entomologist* **83** 324-334.

Tomkins, J. & Simmons, L.W. (2000) Sperm competition games played by dimorphic male beetles: fertilization gains with equal mating access. *Proceedings of Royal Society of London B* **267** 1547-1553.

Werren, J.H., Zhang, W. & Guo, L.R. (1995) Evolution and phylogeny of *Wolbachia:* reproductive parasites of arthropods. *Proceedings of Royal Society of London B* **261** 55-71.

Weston, P. & Diaz, M. (2005) Thermal requirements and development of immature stages of *Viburnum* leaf beetle, *Pyrrhalta viburni* (Paykull) (Coleoptera: Chrysomelidae). *Environmental Entomology* **34** 985-989.

Williams, F.X. (1917) Notes on the life-history of some North American Lampyridae. *Journal New York Entomology Society* **25** 11-13.

Wong, C.H. (2009) Firefly watching and conservation in involving local communities in Malaysia. In: *Diversity and Conservation of Fireflies. Proceedings of the International Symposium on Diversity and Conservation of Fireflies, Queen Sirikit Botanic Garden, Chiang Mai, Thailand, 26-30 August 2008* (Ed. by Napompeth, B.), pp. 141-148. Chiang Mai: Queen Sirikit Botanic Garden.

Lampyrid 2012. **2**:72-79

The life history of a species of *Lamprigera* (Coleoptera, Lampyridae): results of a preliminary study on laboratory rearing of fireflies at Queen Sirikit Botanic Garden, Chiang Mai, Thailand

SOMYOT SILALOM[1*], ANCHANA THANCHAROEN[2] & SUYANEE VESSABUTR[1]

[1] *Queen Sirikit Botanic Garden, 100 Mae Rim-Samerng Road, Mae Ram, Mae Rim, Chiang Mai, 50180, Thailand.*
[2] *Department of Entomology, Kasetsart University, Ngamwongwan Road, Bangkaen, Bangkok, 10900, Thailand.*
**Email: Jeckky@hotmail.com*

ABSTRACT The life cycle of an unidentified species of *Lamprigera* was studied in Chiang Mai province, Thailand, based on ninety six individuals monitored at a maintained temperature of 24 ± 3°C and relative humidity of 80-90 % with a 12 h photoperiod. Each larva was separately reared in a transparent box containing a soil mixture. *Cryptozona siamensis*, a terrestrial snail, was used as the food source for the larvae. Females underwent five instars, while males underwent only three. After the third instar larval males pupated, but the larval period was about the same for both males and females because males remained in the third instar for a long duration. The complete life cycle was estimated to take approximately one year. Average pronotal widths from the first to fifth instars were 5.46 ± 0.18, 6.99 ± 0.53, 8.83 ± 0.73, 11.55 ± 1.51, and 13.40 ± 2.22 mm, respectively. First instar larvae occurred from January to March, and other instars occurred within one to three months of the occurrence of first instar larvae. Pupation occurred in September. Adult females lived two months and had a life span double that of adult males, which lived approximately one month. The diameter of fertilised eggs was approximately 5.56 ± 0.66 mm, with a size increase before hatching. The incubation period was approximately eighty days.

Keywords: Terrestrial firefly, *Lamprigera*, life cycle, *Cryptozona siamensis*, Thailand

INTRODUCTION

Thailand has a rich diversity of fireflies, but little is known about them other than their distribution. The life cycle of *Pteroptyx malaccae*, a semi-aquatic species, has been studied (Loomboot et al. 2007) as well as the life cycle of *Luciola aquatilis*, a freshwater species (Thancharoen et al. 2007) and *Luciola leii* (Fu et al. 2006). Less understood are the terrestrial firefly species, some of which undergo a life cycle very different from the aquatic species. For example, the adult females of terrestrial *Lamprigera* species are larviform, whereas adult females of *Pteroptyx* and *Luciola* are not. Adult males of the former are also much smaller than the adult female (approx. 1 cm and 6–10 cm respectively).

Knowledge of insect life cycles is important in ecological applications. More than 700 articles associated with insect life cycle investigation have been written (Mendez & Resh 2007). Several studies determine the number of generations per year (voltinism) through field observations, such as in the caddisflies *Frenesia missa, Pseudostenophylax uniformis* and *Neophylax concinnus* (Mathis 1999), and *Ugandatrichia maliwan* (Thani & Chantaramongkol 1999). These insects are categorised as univoltine, taking one year to complete a life cycle. The firefly species *Pyrocoelia praetexta* (Sommitr 2004), *P. tonkinensis* (Sommitr 2009) and *Luciola ficta* (Ho et al. 2009) were investigated under indoor laboratory conditions and described as univoltine. However, some species such as *Pteroptyx malaccae* have been deemed to be multivoltine, producing more than two generations per year (Lomboot et al. 2007).

Fireflies in the genus *Lamprigera* are commonly called 'giant fireflies' in Thailand and are easily recognised by their large and bright flash. The present paper presents preliminary results on the life cycle of a terrestrial firefly species from the genus *Lamprigera*. This species looks similar to *Lamprigera tenebrosus* reported by Tittayavan (2009). They both exhibit black tergites with whitish-yellow lateral bands along all tergite margins.

The study was conducted at the Queen Sirikit Botanic Garden (QSBG) in Chiang Mai, Thailand with the objective of developing a protocol for laboratory breeding of this firefly species. Amidst rising concerns that firefly populations are decreasing in Thailand, research on *in-situ* conservation and captive breeding of selected firefly species has been carried out for about a decade under the patronage of Her Majesty Queen Sirikit of Thailand.

METHODS

Collection sites

Lamprigera larvae reared in this study were collected from the Rock Garden and Pine Garden within Queen Sirikit Botanic Garden, as well as from Phanokkok, and the Parkear Training Centre (PTC), all in Chiang Mai, Thailand. The Rock Garden is an outdoor site exhibiting lithophytes and is exposed to the sun, thus having dry top soil. The Pine Garden contains mature pine trees and cycads that shade the garden floor, and the top soil here is moist with grass and leaf litter cover. Phanokkok is located in a mountainous area where intensive agriculture is carried out by hill tribe villagers. The Parkear Training Centre is located on a hillside surrounded by dry evergreen forest with streams that flow only during the rainy season and are otherwise dry year-round. Here, the forest floor is shaded by dense tree canopies and the top soil is covered with leaf litter. Most of the larvae were collected from the Rock Garden and Pine Garden (23 and 22 larvae, respectively), while Phanokkok and the PTC yielded one and two larvae, respectively. An extra male adult was obtained from a site in Moungkum, which is located near to the Phanokkok site.

Laboratory study

The study was conducted from January 2009 to March 2010 in the Firefly Breeding Laboratory of Queen Sirikit Botanic Garden, with room temperature at 24 ± 3 °C and relative humidity between 80-90%. Forty eight field-collected larvae in a range of instars were reared in individual containers. Early instar larvae were kept in smaller round containers (8.5 cm diameter, 4 cm high) while later instars were kept in larger rectangular containers (28×19×15 cm, L×W×H). Wire mesh covering a 4.5 cm diameter hole cut out from the lid of the round containers provided ventilation. For the rectangular containers, there were three such holes on each side of the container. The bottom of the containers was filled with approximately an inch of soil mixture comprising six parts peat moss and one part charcoal chips. A terrestrial snail, *Cryptozona siamensis*, was provided as larval prey. One live snail was placed into each rearing chamber. Dead snails were replaced with live ones throughout the study. The following data were observed and recorded: 1) pronotal width (mm) after each moult, 2) the duration of instars (stadia), 3) duration of the pupal stage, 4) sex of the emerged adult, 5) longevity of the adult, 6) number of eggs laid and hatching success. Pronotal width was used to represent larval size in preference to head capsule width to avoid injuring the heads of the larvae, which are often retracted.

Whenever possible, adult males and females that were obtained were mated for about a week, with one pair per rectangular container (dimensions as above). For example, an emerged female from the Pine Garden site was mated with a male collected from Moungkum. A few days after mating, the female started to lay eggs and the first offspring were reared. A second female to emerge from the Phanokkok site was then mated with the same male from Moungkum. A third female obtained from the Rock Garden was mated with a male that emerged from the Pine Garden. Each newly hatched larva was reared in the same way as the parent fireflies.

RESULTS

Of the forty eight field-collected larvae, six adult females and three adult males were reared out. Laboratory rearing to the adult stage was successful only for larvae collected in the final instars before pupation, while larvae collected in their first and second instars were not successfully reared to adults. Males went through three instars before pupating (as exarate pupae) while females passed through five instars. Most of the adults emerged from late September to early October, but the Parkear

Stage	Female			Male			F1 Progeny (both sexes)		
	Mean ± SD	Range	N	Mean ± SD	Range	N	Mean ± SD	Range	N
Egg	-			-			90.7 ± 5.8	84 - 106	47
Instar I	-			-			50.3 ± 10.5	35 - 73	37
Instar II	-			-			33.9 ± 7.8	19 - 51	18
Instar III	-			105.3 ± 26.6	91 - 136	3	10.0	10 - 10.	1
Instar IV	67.0	67-67	1	N/A			-		
Instar V	87.6 ± 31	50 - 135	5	N/A			-		
Pupa	11.5 ± 0.5	11 - 12.	6	25.3 ± 2.9	22 - 27	3	-		
Adult	38.7 ± 21.7	17 - 71	6	30.3 ± 2.3	29 - 33	3	-		

Table 1 Developmental period and longevity (days) of *Lamprigera* sp. based on larvae collected from various locations in Chiang Mai, Thailand, and the F1 progeny of reared adults. No F1 offspring survived past instar IV. N/A = Not applicable.

Training Centre specimens, which were found later, emerged in December.

Five females mated and all laid eggs, the mean number of eggs per female being 24.4 ± 10.8 (range 14–39 mm). Females began to lay eggs a few days after mating. Five to eighteen eggs were laid per day over a period of four days or less. Mean egg size was 5.6 ± 0.7 mm (range 4.2–6.8 mm, n = 122). Eggs laid by one of the mated females did not hatch, while 14-100 % of the eggs laid by the four other females hatched. Mean incubation period was 90.7 days (Table 1). The unmated female laid 25 eggs that did not hatch.

The first egg cluster was laid by the mated females a few days after copulation. Five to eighteen eggs per cluster were laid in a day over a period of four days or less. Newly laid eggs took approximately 90 days for incubation and first measured approx. 3 mm in diameter growing to 5 mm in diameter before hatching, reflecting that the larva was growing in size inside the egg membrane.

Five of the F1 larval progeny went missing in the first instar and were presumed dead. Cumulative mortality from the first to fourth instar was 12, 57, 98 and 100%, respectively (n = 42). It could be deduced that the sole larva that moulted to the fourth instar and subsequently died was female, but the other larvae could not be sexed.

Mean size of field-collected larvae as measured by pronotal width differed significantly between instars ($F_{4,37} = 26.4$, $P < 0.001$; Table 2). For females in instars

Instar	Female			Male			F1 Progeny (both sexes)		
	Mean ± SD	Range	N	Mean ± SD	Range	N	Mean ± SD	Range	N
I	-			-			5.5 ± 0.1	5.1 - 5.9	37
II	6.9 ± 0.4	6.3 - 7.3	6	-			6.9 ± 0.4	6.3 - 8.5	37
III	9.1 ± 0.7	7.7 - 9.7	9	8.0 ± 0.2	7.7 - 8.1	3	8.5 ± 0.4	7.3 - 9.1	18
IV	11.7 ± 1.2	9.3 - 13.1	10	na			9.3	9.3 - 9.3	1
V	13.4 ± 2.3	10.0 - 16.3	10	na			-		

Table 2 Pronotal width (mm) of *Lamprigera* sp. collected from various locations in Chiang Mai, Thailand and reared to adults, and the F1 progeny. No F1 offspring survived past instar IV. na = Not applicable.

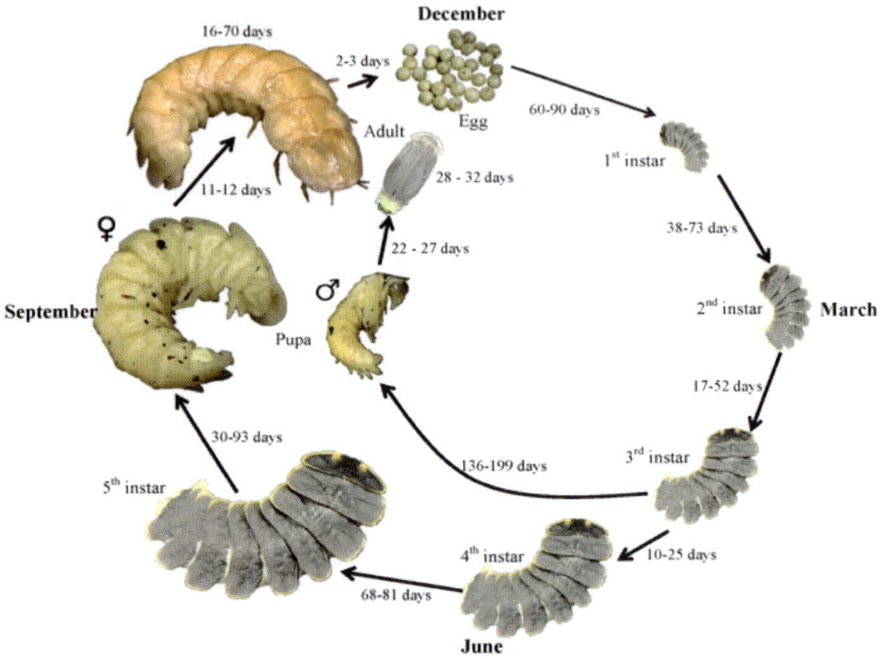

Figure 1 Life cycle of *Lamprigera* sp. observed in the laboratory at 24°C and 70 – 80% relative humidity.

II to V, each instar was significantly different from the instar that followed (Fisher's pairwise comparisons, simultaneous confidence level = 72.7%), but the size of third instar males was not significantly different from that of second and third instar females. Similarly, the first instar larval progeny were smaller than that of the succeeding instar. These offspring were comparable in size with field-collected larvae of the same instar.

Considering the results obtained for instar IV onwards for the field-collected samples, and for instar I to III of their offspring, we conclude that the *Lamprigera* species studied is univoltine, i.e., it produces one brood per year, with the adults emerging in late September (Fig. 1). The first two stadia appear to be the same for both males and females. However, males then undergo a long third stadium (91–136 days) while females pass through two more instars before pupating at around the same time.

DISCUSSION

In order to develop a protocol for laboratory breeding of firefly species, it is necessary to know the life cycle of fireflies such as the giant *Lamprigera*. This study was undertaken with the assumption that the life cycle of *Lamprigera* can be determined under controlled conditions. The *Lamprigera* species selected for this study was chosen for its large body size, which makes observation of its life cycle easier.

To determine the life cycle of the *Lamprigera* sp. in this study it was necessary to combine field collected data with laboratory collected data. The laboratory conditions maintained were without the use of air conditioning, so as to more closely reflect the natural conditions of their habitat. Chiang Mai's hot season peaks in April with temperatures reaching 34.5 °C in April 2010 for 5 days. During this peak in hot weather, larvae became weak in the laboratory then stopped feeding on the snails and died. However, the life cycle of *Lamprigera* sp. was still determined based on field collected samples combined with information obtained from larvae hatched under laboratory conditions.

At the beginning of field collection it was not possible to determine the instar of the collected immatures due to the lack of prior research conducted on this species. Then information was obtained for the first, second and third instars from laboratory breeding. It then became possible to arrange field collected immatures into instars based on their pronotal widths. The *Lamprigera* sp. took one year to complete its metamorphosis and is therefore univoltine. Three quarters of the entire life span was spent in immature stages. The remaining quarter of their life span was spent in the adult stage.

Under laboratory conditions, larvae normally fed in the morning. The head and thoracic segments were inserted into the opening of the snail's shell. Under natural conditions, prey were probably hunted during the night. Specimens were collected only at night. We concluded from this that the larvae of *Lamprigera* sp. are nocturnal, being active and hunting only during the night. It was also observed that larvae never left their prey even when they were disturbed. Whenever larvae were disturbed, they stopped feeding for a while and then resumed.

After feeding, larvae preferred to stay inside their chamber under the soil surface. Forelegs were used to burrow into the soil. A chamber was established under the soil surface. The anterior margin of the pronotum was used a lot for chamber construction. The chambers were globular in shape, suitable for a C-shaped orientation of larvae when in a resting position.

We also investigated the glowing behaviour of larvae. The larvae did not exhibit a specific pattern of flashes. However, from the observation, we found that larvae normally emitted a green flash whenever they were disturbed. It is possible that green light is emitted as a mode of defense.

The study provided further information about the different development periods of male and female *Lamprigera* sp. The laboratory breeding showed that sexes of *Lamprigera* can be differentiated in the third instar. The male passes through only three instars while the female undergoes five instars. The male goes into pupation after the third instar. However, both male and female larvae hunt and feed on the same available terrestrial snails. It is likely that the longer period of development of the female larva is a consequence of its nutritional preparation for oogenesis (Thancharoen 2007) and that the increase in body size is in preparation for the large oocyte mass.

This study not only presents the life cycle of *Lamprigera* sp. but also assists in differentiating this species of unknown taxonomic status from *Lamprigera tenebrosus*, which is distinguished from the former by an orange pronotum and abdominal tergites 6-8. In addition, almost the entire tergal surface of this *Lamprigera* species is black except the tergal plate margins, which have whitish yellow bands. The study also indicates that the habitat requirements of this *Lamprigera* sp. are complex. Further study is needed on this aspect.

ACKNOWLEDGEMENTS

We would like to extend our appreciation to the Queen Sirikit Botanic Garden (QSBG) for financial support. We are also grateful to QSBG entomology staff and others who helped in the field sampling. A special thanks is due to Mr Sutinan Duangdang for his exceptional monitoring of the laboratory rearing experiments.

REFERENCES

Fu, X.H., Ohba, N., Zhang, Y. & Lei, C.L. (2006) A rearing apparatus and diet for the aquatic firefly *Luciola leii* (Coleoptera: Lampyridae). *Canadian Entomologist* **138** 399-406.

Ho, J.Z., Chiang, P.H. & Yang, P.S. (2009) Life history traits of the aquatic firefly, *Luciola ficta* Olivier (Coleoptera: Lampyridae). In: *Diversity and Conservation of Fireflies. Proceedings of the International Symposium on Diversity and Conservation of Fireflies, Queen Sirikit Botanic Garden, Chiang Mai, Thailand, 26-30 August 2008* (Ed. by Napompeth, B.), pp. 100-121. Chiang Mai: Queen Sirikit Botanic Garden.

Loomboot, S., Jamornmarn, S., Chunkao, K. & Chongrattanameteekul, W. (2007) The biology and rearing of firefly *Pteroptyx malaccae* Gorham. *Environment and Natural Resources Journal* **5** 35-43.

Mathis, M.L. (1999) Life histories of three limnephiloid caddisflies (Trichoptera: Limnephilidae, Uenoidae) inhabiting a temporary spring in the Ozark Mountains USA. In: *Proceeding of the 9th International Symposium on Trichoptera, Chiang Mai, Thailand, 5-10 January 1998* (Ed. by Malicky, H. & Chantaramongkol, P.), pp. 217-226. Chiang Mai: Faculty of Science, Chiang Mai University.

Mendez, P.K. & Resh, V.H. (2007) What aspects of the life history of Trichoptera have been studied? In: *Proceeding of the 12th International Symposium on Trichoptera, Chiang Mai, Thailand, 18-22 June 2006* (Ed. by Bueno-Soria, J.R. & Armitage, B.), pp. 191-195. The Caddis Press, Ohio.

Sommitr, W. (2004) *Population Ecology of the Firefly, Pyrocoelia praetexta Olivier in Dok Toh Botanic Garden*. M.Sc. Thesis. Chiang Mai University. 46 pp.

Sommitr, W. (2009) Rearing of some terrestrial firefly species. In: *Diversity and Conservation of Fireflies. Proceedings of the International Symposium on Diversity and Conservation of Fireflies, Queen Sirikit Botanic Garden, Chiang Mai, Thailand, 26-30 August 2008* (Ed. by Napompeth, B.), pp. 92-99. Chiang Mai: Queen Sirikit Botanic Garden.

Thancharoen, A. (2007) *The Biology and Mating Behavior of an Aquatic Firefly Species, Luciola aquatilis sp. nov. Thancharoen (Coleoptera: Lampyridae)*. Ph.D. Thesis. Mahidol University. 105 pp.

Thancharoen, A., Ballantyne, L.A., Branham, M.A. & Jeng, M.-L. (2007) Description of *Luciola aquatilis* sp. nov., a new aquatic firefly (Coleoptera: Lampyridae: Luciolinae) from Thailand. *Zootaxa* **1611** 55-62.

Thani, I. & Chantaramongkol, P. (1999) Life history of *Ugandatrichia maliwan* (Trichoptera: Hydroptilidae) in Mae Klang stream, Doi Inthanon range, northern Thailand. In: *Proceedings of the 9th International Symposium on Trichoptera, Chiang Mai, Thailand, 5-10 January 1998* (Ed. by Malicky, H. & Chantaramongkol, P.), pp. 411-413. Chiang Mai: Faculty of Science, Chiang Mai University.

Tittayavan, M. (2009) Introduction to fireflies in Queen Sirikit Botanic Garden (QSBG). In: *Diversity and Conservation of Fireflies. Proceedings of the International Symposium on Diversity and Conservation of Fireflies, Queen Sirikit Botanic Garden, Chiang Mai, Thailand, 26-30 August 2008* (Ed. by Napompeth, B.), pp. 122-125. Chiang Mai: Queen Sirikit Botanic Garden.

Lampyrid 2012. **2**:80-86

Observations on the external morphology and behaviour of *Pteroptyx tener* and *Pteroptyx bearni* in Malaysia

Nobuyoshi Ohba[1]* & Choong Hay Wong[2]

[1]*Ohba Firefly Institute, 4-1-12-204 Maborikaigan, Yokosuka City. 239-0801 Japan.*
[2]*Malaysian Nature Society, JKR 641, Jalan Kelantan, Bukit Persekutuan, 50480 Kuala Lumpur, Malaysia.*
**Email: qgb00523@nifty.ne.jp*

ABSTRACT Colonies of *Pteroptyx tener* and smaller colonies of *Pteroptyx bearni* can be found along the Kerteh River in Terengganu, while only the latter is found 42 km to the south at Cherating River, Pahang. Aspects of the external morphology, habitat and behaviour of these firefly species were studied at these two locations. At the Kerteh River, *P. tener* was found in the brackish region while *P. bearni* was found in more saline regions near the estuary, with some overlap in the area of occurrence. Body length of *P. bearni* is smaller than *P. tener* and they also differ in colouration. *P. tener* flashes synchronously, while *P. bearni* does not flash synchronously. The two species also differ in flash communication behaviour, including duration and intensity. *P. bearni* males produce a brighter light than *P. tener*. Both firefly species display on vegetation commonly found at their display site. *P. tener* and *P. bearni* sometimes share the same trees when both species were sympatric. Colony size and density were much lower in *P. bearni* compared to *P. tener*.

Keywords: Firefly, morphology, ecology, *Pteroptyx tener, Pteroptyx bearni*, Malaysia

INTRODUCTION

Malaysia is known to harbour the unique congregating fireflies of the genus *Pteroptyx* Olivier 1902 (Lampyridae: Luciolinae). The Malaysian congregating fireflies have been reported primarily displaying on mangrove trees and riparian trees along the tidal rivers (high salinity, brackish, and acidic black water). There are two main types of congregating fireflies, the synchronous and the non-synchronous (Ohba 1983; Ohba & Sim 1994; Ohba & Wong 2004; MNS 1996; Wong & Yeap 2012). The study was conducted at the Kerteh River and Cherating River on the east coast of Peninsular Malaysia. Both *Pteroptyx tener* and *P. bearni* are found at the Kerteh River, while a third species, which occurs in very small numbers high up on 10 metre-high trees, was not identified (Wong & Yeap 2012). This is possibly a first record for *P. bearni*, in Peninsular Malaysia, although it is known from Singapore and Sarawak. However the identity of *P. bearni* is still under investigation by Ballantyne. It was first described from Singapore in 1909 by Olivier (Ballantyne & McLean 1970). This study was based on the colony of *P. tener* found by Rozaimi Jamaludin et al. (2007) at Kampung Ranggun, Kerteh. The Kerteh River had a detailed survey done to establish the congregating firefly zones (CFZs) by GPS mapping of firefly colonies, land use and threats during the dry season. Identification of the display trees was done by Wan Faridah and Wong.

A chance visit to Cherating River by the authors in November 2008 led to the discovery of *P. bearni* resting on *Rhizophora* sp. leaves (a true mangrove tree) in the daytime. In October 2009, a CFZ survey was conducted in Cherating by one of the authors (Wong & Yeap 2012) and this location was found to be suitable for studying *P. bearni*. This river supports a single firefly species. Both of these species were observed by the authors to be present in Chukai River in Terengganu (a river between the Kerteh River and Cherating River) during the 2008 study.

The authors continued their earlier study on *P. tener* (Ohba & Wong 2004), and conducted a preliminary study of *P. bearni* at Kerteh River and Cherating River in November 2008 during the monsoon season. The study looks into the habitat range and reconfirms the preference of each species for either a more saline or brackish regions (Wong & Yeap 2012). The external morphology, behaviour and habitat of both species are discussed.

METHODS

Study areas

The study areas were along the estuaries of the Kerteh River (N 4° 32', E 103° 26'), in Terengganu, and the Cherating River (4° 07', E 103° 23'), 42 km south in Pahang. Both rivers flow to the South China Sea. The Kerteh River (Fig. 1A) flows through a mangrove forest that is one of the most extensive in the East Coast, currently with an unprotected status. The river flows from Bukit Patang (629 m) in the Rasau Kerteh Forest Reserve, a lowland dipterocarp forest, before meandering through peat swamp and mangroves to the South China Sea (DWNP 1987). The river is relatively undisturbed except near the petroleum industrial area, agriculture areas and villages. The upper estuarine regions still have stands of tall, old *Sonneratia caseolaris* trees and unharvested *Nypa fruticans*. The succession of vegetation along the Kerteh River is evident with stands of *Rhizophora, Bruigeira, Ceriops, Xylocarpus* and *Aviccenia* downstream, and progressively upstream the true mangrove trees are replaced by *Nypa, Sonneratia, Talipariti* (Hibiscus), *Barringtonia, Gluta, Pandanus, Cerbera, Excoecaria* and various other associated trees. This upper zone is the *P. tener* study area. A rare mangrove plant, *Brownlowia argentata* is commonly found here. The difference in the water level is between 1 to 2 m, caused by tidal change. The flow of the river changes direction whenever there is a high tide, and the salinity intrusion goes more than 13 km upstream. Wildlife is relatively abundant.

Figure 1 The Kerteh River (A) and Cherating River (B) with various display trees.

The Cherating River (Fig. 1B) flows from the Jabor Forest Reserve with an average altitude of 71 m asl, meandering through the Cherating Mangrove Forest Reserve, which is a good example of estuarine mangrove on the East Coast of Peninsular Malaysia (DWNP 1987). The short river flows east from the hills before turning south west and running parallel to the coast before reaching the sea. The *P. bearni* study area was within the mangroves. The vegetation is dominated by *Rhizophora* and *Bruguiera*. Other species include *Ceriops, Avicennia, Lumnitzera, Xylocarpus,* and *Acanthus* (DWNP 1987). Very few *Nypa* palms were observed. The mean tidal variation is 1.2 m (DWNP 1987). The riverbank is relatively pristine and undisturbed. The flow is calm and the river narrows at the upper stream and widens at the estuary, and both are inaccessible during low tide. The river water is peaty.

The external morphology of the fireflies was photographed and measured using a binocular microscope. Photographs of both habitats were taken, and data recorded about the observation sites. Flash patterns were recorded by video camera (Richow Caplio GX100) with a light intensifier tube, and infra red light and ambient temperatures were recorded concurrently.

RESULTS

Morphological observations

The principle morphological feature differentiating the two species in this region was the colour of the pronotum (Fig. 2). In *P. bearni* there is a greater orange pigmentation of the pronotum than in *P. tener*; this is evident for both males and females. The majority of other morphological features were found to be the same for both species. Male genitalia was also found to be highly similar in appearance.

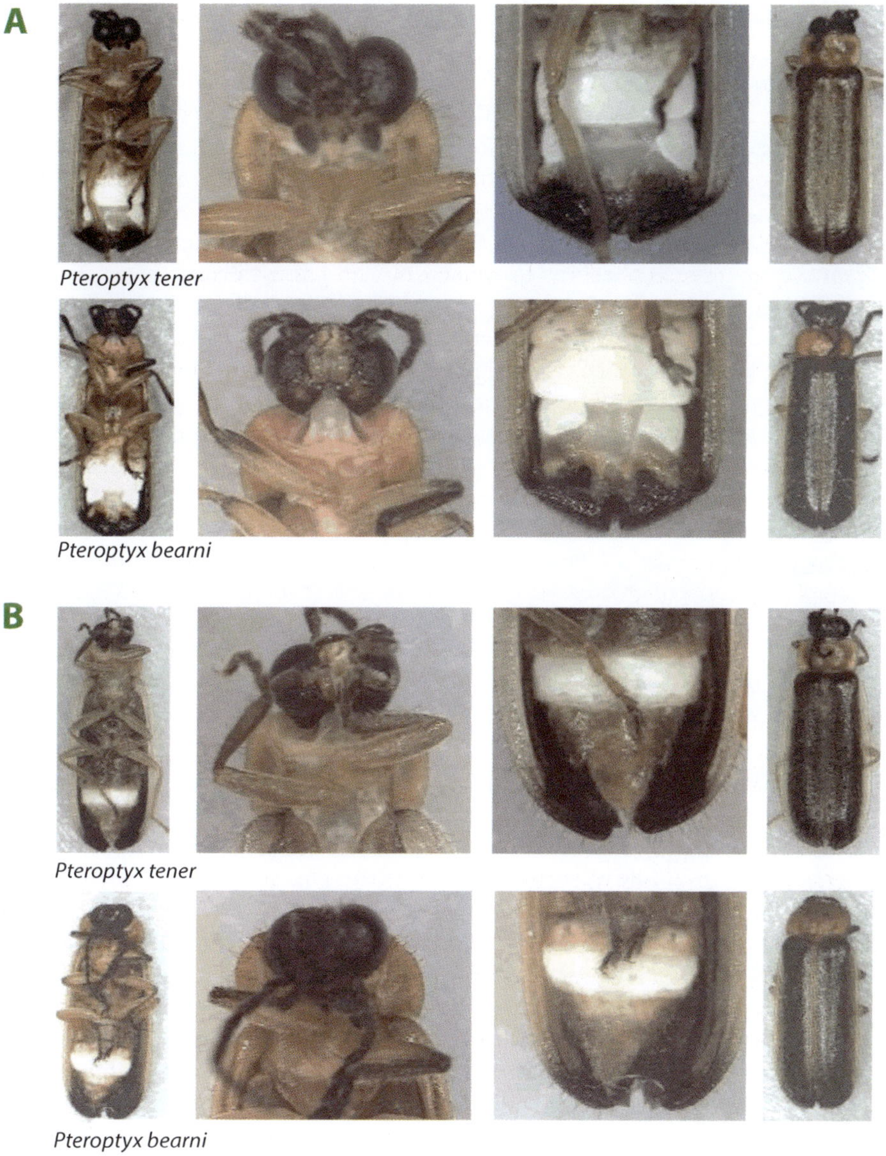

Figure 2 Comparison of external features between **A.** *P. tener* and *P. bearni* adult males and **B.** *P. tener* and *P. bearni* adult females.

Figure 3 *P. bearni* resting on the leaf of a mangrove tree during the day.

Behavioural observations

At night *P. tener* was sometimes observed on *Nypa* palm leaves, but in small numbers. *P. bearni* were observed to be less active than *P. tener* at night although *P. bearni* males were more active than their females. During the day, *P. bearni* rested on the upper side of leaf surfaces but was not seen on the underside or the stem (observed during overcast skies and humid days) (Fig. 3).

The female's flash pattern looked similar to the males in *P. tener* and *P. bearni*. It was difficult to distinguish between the two sexes in the field for both species. The males of *P. tener* perched on leaves and synchronised their flashing, creating regular flash patterns, but the males of *P. bearni* did not appear to synchronise their flashing.

The *P. bearni* male flash pattern had various peaks (1-3) and a shoulder every 1 to 2 seconds or longer in interval (Fig. 4B). The flash cycle was not regular or in synchrony, unlike that of *P. tener* (Fig. 4A). The former will emit light every 0.5-1.0 s when put in a transparent plastic bag. The luminescence cycle is shorter than in the field. The light is also weaker, but over time the intensity increased. Two male flash patterns were observed, one of them a single steady flash and the other composed of flickering flashes.

The flash pattern of non-synchronous *P. bearni* may change over time. Determining this needs more stationary observation of a chosen

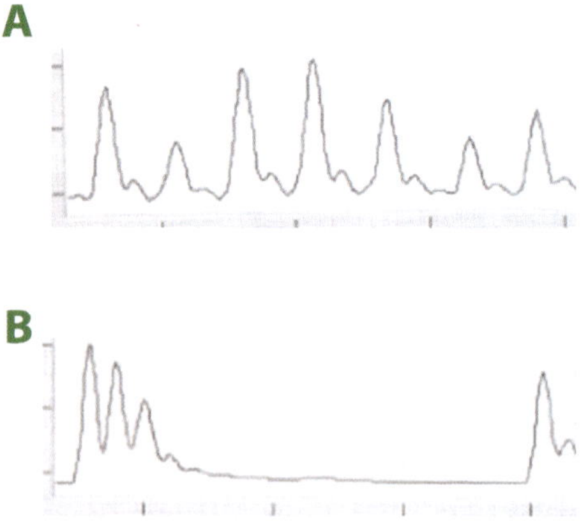

Figure 4 Adult male flash pattern; A. *Pteroptyx tener*; B. *P. bearni*.

display tree, but time was insufficient during this study.

DISCUSSION

The preferences of *P. bearni* for lower estuarine habitats and *P. tener* for upper estuarine, brackish-water habitats were reconfirmed by their distribution along a single river. The habitats of *P. bearni* and *P. tener* are largely isolated from each other. The display tree is usually different, but sometimes their habitats do overlap. Unlike *P. bearni*, *P. tener* is a congregating, synchronously-flashing firefly, found in greater numbers per tree and in a greater range of trees. A further difference is that *P. bearni* is found higher up the tree than *P. tener*. Sizes of both species are very similar but *P. bearni* has darker elytra and a more orange pronotum. Both males and females can be differentiated morphologically and behaviourally. The similarity in male genitalia makes introgression between *P. tener* and *P. bearni* a distinct possibility. However, both habitat preferences and flash patterns are different, and consequently both species are considered isolated from each other and valid species.

In the evening, *P. tener* will congregate, and after sunset synchronisation begins. The males are active on the trees. Synchronous flashing plays a major role for sexual communication and mating. After midnight when the peak copulation period is over the females actively flash and at a higher density. This curious behavior has yet to be explained.

ACKNOWLEDGEMENTS

We would like to thank MNS colleagues Wan Azmadi, Ahmad Azri Alias, Zaid Hamzah, Mohd Rozaimi and Shanmugaraj, as well as Ms Juhaida (Universiti Kebangsaan Malaysia), Wan Faridah Akmal Wan Jusoh (Universiti Pertanian Malaysia), Nur Wahida (Universiti Sabah Malaysia) for helping in the survey. We also thank Pak Hashim (EcoCare) and Hafiz, both knowledgeable boatmen. Other individuals not mentioned here are duly acknowledged. This work was undertaken with the support of The Mohamed Bin Zayed Species Conservation Fund, project no. 0925338.

REFERENCES

Ballantyne L.A. & McLean M.R. (1970) Revisional studies on the firefly genus *Pteroptyx* Olivier (Coleoptera: Lampyridae: Luciolinae: Luciolini). *Transactions of the American Entomological Society* **96** 223-305.

DWNP (Department of Wildlife and National Parks). (1987) *Malaysian Wetland Directory*. Department of Wildlife and National Parks, Peninsular Malaysia.

MNS (Malaysian Nature Society). (1996) *Strategies for Conservation and Tourism Development at Sungai Selangor Riparian Forest. Final Report Submitted to the Selangor State Government. Kuala Lumpur.* Malaysian Nature Society.

OHBA, N. (1983) Studies on the communication system of Japanese fireflies. *Science Report of the Yokosuka City Museum* **30** 1-62.

OHBA, N. & SIM, S.H. (1994) The morphology and life cycle of *Pteroptyx valida* (Coleoptera: Lampyridae) in Singapore. *Science Report of the Yokosuka City Museum* **42** 1-11.

OHBA, N. & WONG, C.H. (2004) External morphology and ecological study of the firefly, *Pteroptyx tener* at Kampong Kuantan, Selangor, Malaysia. *Science Report of the Yokosuka City Museum* **51** 1-33.

ROZAIMI JAMALUDIN, M., WONG, C.H., NUR WAHIDA, Y. & SHANMUGARAJ, S.A. (2007) *Feasibility Study for a Firefly Conservation Project Involving the Local Community at Sungai Kertih, Terengganu.* Malaysian Nature Society. 33 pp.

WONG, C.H. & YEAP, C.A. (2012) Conservation of Congregating Firefly Zones (CFZs) in Peninsular Malaysia. In: *Firefly Conservation: From Science to Practice. Proceedings of the Second International Firefly Symposium, Selangor, Malaysia, 2-5 August 2010* (Ed. by Kirton, L.G., Lim, G.T. & Day, J.C.). *Lampyrid* **2** 174-187.

Lampyrid 2012. **2**:87-98

Male aggregation behaviour in the fireflies *Photinus ignitus* and *Pteroptyx malaccae*

Christopher Cratsley[1*], Tanthai Prasertkul[2] & Anchana Thancharoen[3]

[1] *Fitchburg State College, 160 Pearl Street, Fitchburg, MA 01420 USA.*
[2] *Mahidol University, Rama VI Road, Rachadhavi, Bangkok 10400, Thailand.*
[3] *Kasetsart University, 50 Phaholyothin Road, Ladyao, Jatujak, Bangkok 10900, Thailand.*

**Email: ccratsley@fitchburgstate.edu*

ABSTRACT Apart from the bioluminescence involved, the remarkable synchronously flashing aggregations formed by males of some *Pteroptyx* species bear little resemblance to the roving male flash behaviour observed in *Photinus* and many other firefly species. However, synchrony has been documented in some *Photinus* species and could be an emergent property of competitive flash behaviour among males at high densities. We used "virtual fireflies" (VFIs), programmable microcomputers driving LEDs, to explore how males aggregate into potentially high density groups in both *Photinus ignitus* and *Pteroptyx malaccae*. We observed the frequency of flashes from roving male *P. ignitus* fireflies at the sites of VFIs programmed to conspecific male and female flash patterns as well as in the absence of flashes in Massachusetts. We found roving males were more frequent and flashed more frequently at sites with conspecific male and female flash patterns. To explore the role of flash synchrony and conspecific signals in male *P. malaccae*, we varied VFI flash treatments each night at two sites along a jetty on the Chao Phraya River in Samut Prakarn, Thailand. These treatments included conspecific and heterospecific synchronous flash patterns as well as conspecific flash patterns produced asynchronously. We found significant differences across treatments in the number of fireflies approaching and synchronising with the virtual fireflies. In each case we observed the greatest number of fireflies approaching, landing near, or synchronising with the synchronous species-specific flash pattern. Our research illustrates that males approach conspecific flash patterns while searching in both *Photinus* and *Pteropyx* fireflies, but that *P. malaccae* males only approach synchronously flashing groups of males. While it remains unclear how synchrony has evolved in *Pteroptyx*, it seems likely that male competition for access to females has played a role in selecting for the evolution of both loose roving aggregation behaviour in *P. ignitus* as well as the more dense stationary leks observed in synchronous *P. malaccae*.

Keywords: Firefly, male aggregation, behaviour, *Pteroptyx malaccae*, *Photinus ignitus*

INTRODUCTION

One of the most impressive of all animal courtship displays is the mass flash synchrony exhibited by several *Pteroptyx* species in Southeast Asia. Males of some *Pteroptyx* species, including *P. malaccae*, congregate in leks and emit synchronised courtship flashes that attract females (Buck 1988; Buck & Buck 1978; Lloyd 1979; Ohba 2004). Synchrony has also been observed in Asiatic *Luciola* (Ohba 1984; Ohba 2004) and *Pygoluciola* (Fu & Ballantyne 2008), and the related phenomena of wave and intermittent synchrony among flying males have been reported in several North and South American fireflies (Buck 1988; Case 1984; Cicero 1983; Copeland & Moiseff 2004; Lloyd 1983; Viviani 2001). These examples of flash synchrony are striking in their contrast to the discontinuous and asynchronous flash displays observed in most roving male fireflies.

In the firefly species with roving males, including North American *Photinus* and *Pyractomena* (Lloyd 1971, 1979), flying males emit spontaneous, bioluminescent, advertisement flashes. Once a female responds by flashing, a courtship dialogue begins in which the male and female exchange flash signals until the male contacts the female and copulation occurs. While female preferences for the temporal characteristics of male courtship signals have been documented in a number of *Photinus* species with roving males (Branham & Greenfield 1996; Cratsley & Lewis 2003; Michaelidis et al. 2006), male competition may also play a critical role in the evolution of these courtship behaviours (Vencl & Carlson 1998; Lewis & Cratsley 2008).

Many authors have reported that firefly mating aggregations show male biased operational sex ratios, particularly early in the mating season (Cratsley & Lewis 2005; Kaufmann 1965; Lewis & Wang 1991; McDermott 1964), which can lead to intense male-male competition for access to females. The differences in male courtship flash-signalling behaviours, from synchronous flashing to asynchronous roving flights, are accompanied by a range of male competitive behaviours (reviewed by Lewis & Cratsley 2008), from examples of female defence in which males guard individual females in both *Photinus* (Lewis & Wang 1991; Wing 1985) and *Pteroptyx* (Wing et al. 1983), to scramble competition, and lek formation. Male competition for access to females may select for similar types of competitive behaviours across different firefly species.

Several male flash-signalling and searching behaviours observed in fireflies may reflect mate competition. As firefly males search for females, they are attracted to ongoing male-female flash dialogues (Lloyd 1979; Papi 1969; Vencl & Carlson 1998). Several types of flash competition have been described between males engaged in dialogues with the same female in both *Photinus* and *Pteroptyx* (Carlson & Copeland 1988; Case 1980; Lloyd 1979). These findings suggest that male fireflies may be highly responsive not only to female signals but to the signals of competing males. These male-male interactions could play important roles both in the scramble competition behaviours of roving *Photinus* males and in the establishment of synchronously flashing *Pteroptyx* leks.

In order to better understand the roles of male-male competition in the search behaviour of North American *Photinus* fireflies and lek formation in *Pteroptyx*, we used Virtual Firefly Instruments (VFIs) in the field to simulate male (and in some cases female) flash signalling behaviour and observed the responses of male *Photinus ignitus* and *Pteroptyx malaccae* fireflies. *Photinus ignitus* fly and produce single flash pulses repeated approximately every 10 seconds, and females respond to these flashes from perches in the vegetation after a species-specific time delay (Lloyd 1966). While males use these female responses to locate and mate with females, we hypothesised that *P. ignitus* males might also respond to and orient towards simulated conspecific male flash signals to compete with other males that may already be engaged in dialogues with females.

In *P. malaccae,* we hypothesised that not only would males respond to and orient towards conspecific male flash signals, but that they would be highly sensitive to the synchrony of these flashes. Several hypotheses for the evolution of synchronous flash behaviour in fireflies have been proposed (Buck 1988; Buck & Buck 1978; Case 1984; Greenfield 2005; Lloyd 1983; Otte & Smiley 1977), including the possibility that it is an adaptation to prevent the same reduction in female response to asynchronous competing signals that has been observed in the *Photinus* species *P. pyralis* (Case 2004) and *P. carolinus* (Moiseff & Copeland 2010). Therefore, we hypothesised that in order to ensure female response, *P. malaccae* males would preferentially join simulated leks flashing synchronously at their species-specific flash rate over simulated synchronous leks flashing at higher or lower flash rates, or asynchronous leks.

METHODS

Virtual Firefly Instrument

The Virtual Firefly Instrument (VFI) is a hand-held, microcontroller-based device that was designed at Fitchburg State University to emulate the signalling behaviour of both male and female fireflies. A photodiode is used to detect the flashing of a nearby firefly (within one metre). The photodiode's analogue signal is amplified, filtered and channelled to the A/D input of a PIC microcontroller. Software in the microcontroller periodically samples the signal from the photosensor until a flash with particular characteristics is detected. When an incoming flash is recognised, or if a flash has been programmed to repeat at regular intervals, a species-appropriate flash is scheduled and made using a pulse-width modulated signal channelled to a yellow LED.

The characteristics of detected flashes required to trigger a response, the response flash, the response delay interval and flash interval are field-programmable. The times and durations of flashes received and transmitted are logged in EEPROM and are retrievable through an RS-232 connection. The hardware for this microcontroller-based instrument is based on the PIC16F88 microcontroller with all hardware functions (power supply, microcontroller, data storage and photometer circuitry) on a single PC Board. Microsoft® Windows based software on a laptop computer was used in the field to program free-running LED flash durations and interflash intervals for VFI flash signalling experiments with *P. ignitus* and *P. malaccae*.

Photinus ignitus male search behaviour

Photinus ignitus adults were observed at Dexter Drumlin in Lancaster Massachusetts, USA on eleven nights between 12 July and 1 August 2005. We used VFIs to establish four field treatments spaced out in a diamond-shaped configuration within a known *P. ignitus* mating habitat, with each VFI located 15 m from the next. The treatments were each assigned to a specific location on a given night and rotated clockwise from one location in the field to the next location on subsequent nights to control for location effects. Each treatment was set up on an arena consisting of a darkly painted, circular piece of plywood (1.0 m in diameter) painted with concentric white rings 5.0 cm apart to allow observers to note where male fireflies landed.

Treatment 1 simulated a responsive female firefly, using a VFI placed in the centre of the arena. The VFI was programmed to spontaneously produce a flash of a duration that mimics the flash of a female *P. ignitus* every 30 seconds. Treatment 2 simulated a spontaneously-flashing male using a VFI placed on a wooden platform 25 cm from the centre of the arena and programmed to produce flashes at a male *P. ignitus* flash duration and inter-flash interval. Treatment 3 combined the first two to simulate a firefly courtship in progress using an elevated, spontaneously flashing male-programmed VFI and the central female-programmed VFI. Treatment 4 was a control consisting of an empty arena.

Observations were conducted over a series of ten minute intervals each night (21:00–23:00 hrs). The four treatments were split between two observers so that each person visited the same two arenas each night. Behaviour of fireflies was recorded when they entered the study area (defined by a 2 m radius around the centre of the arena). The firefly's flash pattern, duration in the area, and position relative to or on the arena were recorded. We compared the number of *P. ignitus* flashes per minute at each of the treatments using a Kruskal-Wallis test because the data was not normally distributed. Individual comparisons between treatments were conducted using the Dunn's procedure.

Pteroptyx malaccae male lek formation

Pteropytx malaccae adults were observed each night from 23–30 November 2008 at two sites along a jetty on the Chao Phraya River in Samut Prakarn, Thailand. The treatments, produced by a linear array of four VFIs spaced at 10cm intervals, were assigned to different time slots within each night and were alternated between the two sites from 20:00–22:00 hrs. The site at which observations were initiated was also alternated on different nights, and the order in which we presented the treatments at each site was rotated to ensure that each treatment was presented first, second, third and fourth in the sequence of nightly observations.

The treatments included one in which all four VFIs flashed synchronously at the species-specific rate observed in the field that night (treatment 1), a second in which the VFIs flashed synchronously at half the rate observed in the field that night (treatment 2), a third in which the VFIs flashed synchronously at twice the rate observed in the field that night (treatment 3), and a fourth in which the VFIs flashed asynchronously, with one VFI at the species-specific rate and the others at 50 ms faster, 50 ms slower and 150 ms slower than the species-specific rate, respectively (treatment 4).

Each treatment was observed for 15 min at each site, and we counted the number of fireflies that approached the VFIs, i.e., that flew above the jetty, that landed on the vegetation where the VFIs were located, and that flashed synchronously with at least one of the VFIs. Data was pooled across the two observation sites. Because the data was not normally distributed, we compared numbers of fireflies engaging in each of these behaviours across treatments using a Kruskal-Wallis test. Individual comparisons between treatments were conducted using the Dunn's procedure.

RESULTS

Photinus ignitus male search behaviour

P. ignitus males were observed flying and flashing above the fields at Dexter Drumlin during the nightly observation periods. The frequency of flashes registered at the arenas over the test periods differed significantly across treatments (Kruskal-Wallis $H = 10.53$, $p < 0.025$), with VFIs receiving more visits (as indicated by flash frequencies) than the control (Fig. 1). The frequency of flashes at the control treatment differed significantly from the frequency at both the simulated courtship treatment (Dunn's $Q = 3.35$, $p < 0.005$) and the simulated male treatment ($Q = 2.77$, $n = 11$, $p < 0.05$). There were no other significant differences between individual VFI treatments (all $p > 0.2$). Over all of our nights of observation, only four fireflies were ever observed flying over the control treatment, while roughly equivalent total numbers of fireflies were observed over the three other treatments (32 at the simulated courtship treatment, 28 at the simulated male treatment, and 25 at the simulated female treatment).

Pteroptyx malaccae male lek formation

Pteroptyx malaccae males synchronously flashing in established leks on mangrove trees along the Chao Phraya River sometimes left their leks and approached the flashing VFIs. There were significant differences across treatments in the number of males approaching the VFIs (Fig. 2, Kruskal-Wallis $H = 12.03$, $p < 0.01$) and synchronising with VFIs (Fig. 3, $H = 12.51$, $p < 0.01$), but not in the number of fireflies

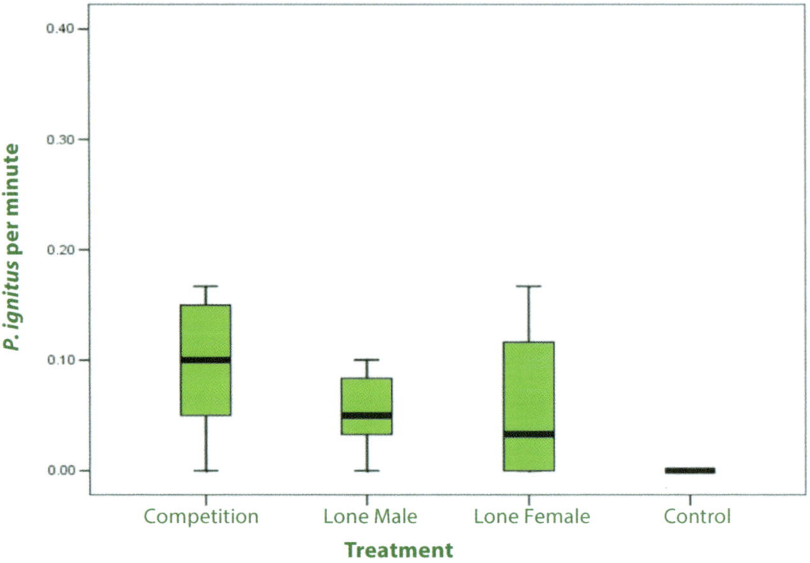

Figure 1 Box plot of the number of flashes of *P. ignitus* observed at sites with VFIs simulating a male and female flash dialogue (competition), a lone flashing male and a lone flashing female, compared against a control site with no VFI.

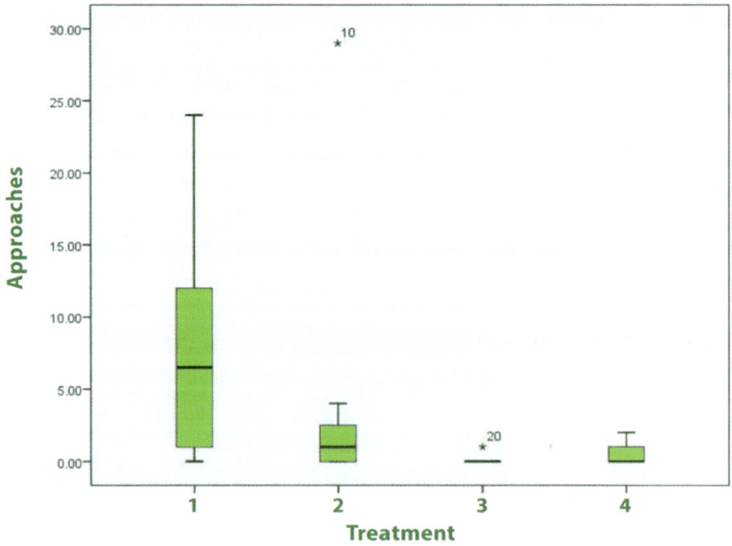

Figure 2 Box plot of number of *P. malaccae* approaching VFIs flashing synchronously at the rate observed in the field (treatment 1), at half the rate (treatment 2), at twice the rate (treatment 3) and asynchronously (treatment 4).

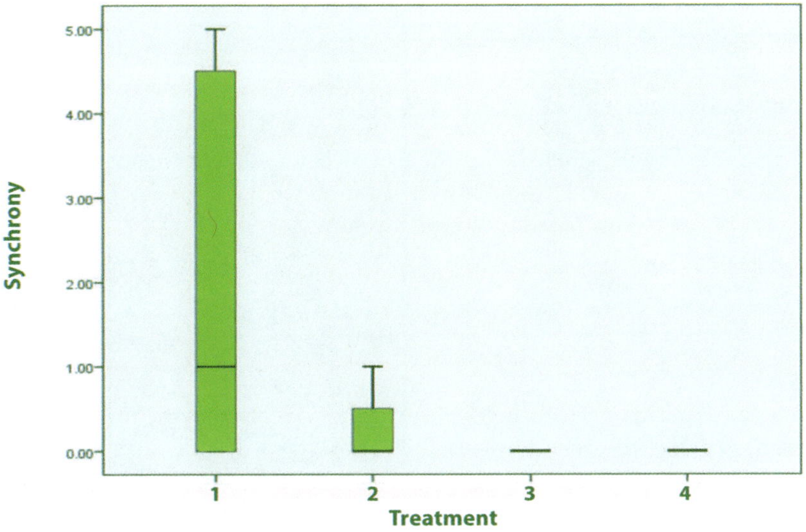

Figure 3 Box plot of number of flashing *P. malaccae* that synchronised with VFIs flashing synchronously at the rate observed in the field (treatment 1), at half the rate (treatment 2), at twice the rate (treatment 3) and asynchronously (treatment 4).

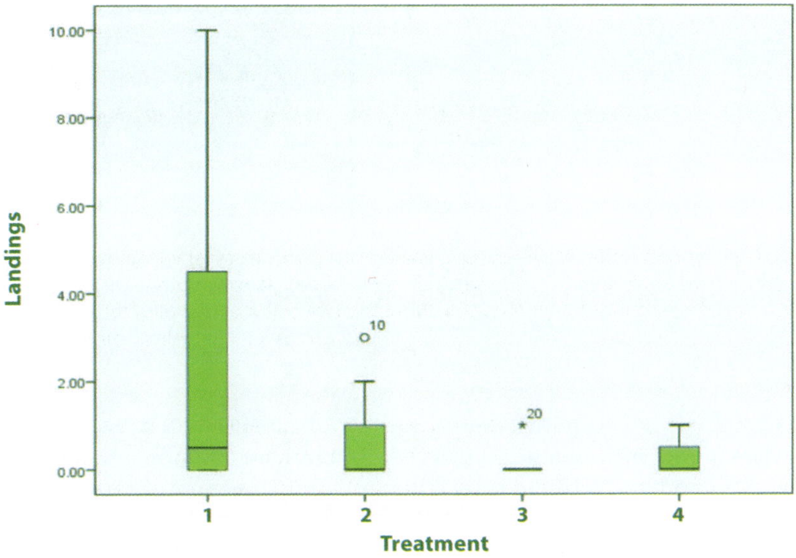

Figure 4 Box plot of number of *P. malaccae* landing near VFIs flashing synchronously at the rate observed in the field (treatment 1), at half the rate (treatment 2), at twice the rate (treatment 3) and asynchronously (treatment 4).

landing near VFIs (Fig. 4, H = 4.05, $p > 0.2$). We observed the greatest number of fireflies approaching, landing near, or synchronising with the VFIs when the treatment with the species-specific, synchronous flash pattern was in use. Significantly more fireflies approached this treatment than approached the treatment producing flashes at twice the normal rate (Dunn's Q = 3.28, $p < 0.01$). In addition, significantly more fireflies synchronised with the flashes in this treatment than with the flashes in either the treatment producing flashes at twice the normal rate or any of the flashes produced by the VFIs in the asynchronous treatment (Q = 2.70, $p < 0.02$ for both). There were no other significant differences between individual VFI treatments (all $p > 0.2$).

DISCUSSION

Our field-based experiments utilising VFIs provide evidence that the males of both *P. ignitus* and *P. malaccae* orient towards and approach conspecific male flashes. In *P. ignitus*, we found no difference in the rate at which males approached simulated female flashes, male flashes or simulated courtships consisting of male and female flashes, suggesting that they orient towards and approach both male and female flashes. *Pteropytx malaccae* males preferentially approached synchronously flashing VFIs flashing at the species-specific flash rate over VFIs producing synchronous flashes at double the species rate. It appears that male *P. malaccae* will leave existing leks, investigate and potentially join a different lek, if males in that lek are flashing synchronously at the appropriate rate.

In both species, the aggregation of males that occurs as they orient towards each other's signals may represent a competitive behaviour that increases their opportunities to find and mate with difficult to locate females under male-biased sex ratios. *Photinus ignitus* and *P. malaccae* females differ in their mating behaviour, with *P. ignitus* females tending to remain stationary on perches (Lloyd 1966; Cratsley & Lewis 2005), while *P. malaccae* females move among mangrove leaves and branches (Buck 1988; Buck & Buck 1978; Lloyd 1979; Ohba 2004). However, females tend to attract multiple, flashing males in both *P. ignitus* (Cratsley & Lewis 2005) and *P. malaccae* (personal observation), creating the potential for males to locate females by seeking out the flashes of other males.

In *P. ignitus*, female responsiveness to male flash signals decreases as the number of nearby signalling males increases (Cratsley & Lewis 2005). This is consistent with lower female responsiveness when sex ratios are male biased in *Photinus aquilonius* and *Photinus marginellus* (Lewis & Wang 1991) and with reductions in female responsiveness when multiple flashes are presented asynchronously in *P. pyralis* (Case 2004) and *P. carolinus* (Moiseff & Copeland 2010). This lower female responsiveness may select for male competition through male aggregation with conspecific flashing males as well as synchronous flashing behaviour in dense male assemblages.

Occurrences of synchronous flashing have been recorded for some *Photinus* species, including *P. pyralis* (Buck 1938; Mast 1912; Rau 1932; Richmond 1930), *P. concisus* (Otte & Smiley 1977), and *P. knulli* (Cicero 1983). In addition, synchronous male flashing has been well documented for *P. carolinus* (Copeland & Moiseff 1995; Copeland et al. 2008; Faust 2010; Moiseff & Copeland 1995, 2010; Moiseff et al. 2001). Given the occurrence of highly competitive love knots in *P. pyralis* and *P. carolinus* (Faust 2004; Maurer 1968; Vencl & Carlson 1998), it is possible that high male densities caused by male aggregation in response to conspecific male flashes are an important factor in the evolution of synchronous flashing behaviour in *Photinus*.

The aggregation behaviour of *P. malaccae* may represent a further step in the evolution of synchronous behaviour, as males preferentially approach and synchronise with leks synchronously flashing at the appropriate species rate, over leks flashing at twice the rate or flashing asynchronously. These findings could provide some insight into the forces shaping the evolution of synchrony in *Pteroptyx*. A variety of hypotheses have been put forward to explain how natural selection might have favoured the evolution of synchronous flash behaviour in *Pteroptyx* (Buck & Buck 1978; Case 1984; Greenfield 2005; Lloyd 1983; Otte & Smiley 1977). In synchronously flashing *P. carolinus* fireflies there is evidence that females respond to a species-specific rhythm and males flash synchronously to preserve this rhythm (rhythm preservation model, Moiseff & Copeland 2010). Our findings of *P. malaccae* male preference to join synchronous leks flashing at the species-appropriate rate are consistent with the rhythm preservation model, but do not rule out other hypotheses.

Testing the many hypotheses for the evolution of flash synchrony in *Pteroptyx* requires a better understanding of the factors influencing individual reproductive success in these displays than is currently available (Lewis & Cratsley 2008). Developing this understanding will involve investigations into female preferences for synchronous and asynchronous flashing in *Pteroptyx*, the outcomes of direct male-male interactions, the potential for postcopulatory male competition and female choice, and the indirect interactions of males leaving and joining leks based on lek size, synchrony, and other photic cues. The recognition that males in very different firefly species such as *P. ignitus* and *P. malaccae* respond to and orient towards conspecific male flashes provides us with a better understanding of one factor that may have driven the evolution of synchrony and other firefly signalling behaviours.

ACKNOWLEDGEMENTS

We would like to thank Fitchburg State University, Kasetsart University, and Mahidol University for helping to support this research. In addition, we thank Martina Rosado, Gina Basinsky, and Dr. Paul Moosman for their assistance conducting field work in Massachusetts, and the Thancharoen family for their hospitality during fieldwork in Thailand. We also thank The Committee on the

Study of Fireflies in Thailand Project, The Queen Sirikit Botanical Gardens, The Malaysian Nature Society and Forest Research Institute Malaysia for their roles in hosting the International Firefly Symposia in 2008 and 2010 that made this international collaboration possible. This research was supported by NSF DBI 0330897, a Marion and Jasper Whiting Foundation Grant, and a Fitchburg State College Ruth Butler Grant.

REFERENCES

BRANHAM, M.A. & GREENFIELD, M.D. (1996) Flashing males win mate success. *Nature* **381** 745–746.

BUCK, J.B. (1938) Synchronous rhythmic flashing of fireflies. *Quarterly Review of Biology* **13** 301–314.

BUCK, J.B. (1988) Synchronous rhythmic flashing of fireflies. II. *Quarterly Review of Biology* **63** 265–289.

BUCK, J.B. & BUCK, E. (1978) Toward a functional interpretation of synchronous flashing by fireflies. *American Naturalist* **112** 471–492.

CARLSON, A.D. & COPELAND, J. (1988) Flash competition in *Photinus macdermotti* fireflies. *Behavioural Ecology and Sociobiology* **22** 271–276.

CASE, J.F. (1980) Courting behaviour in a synchronously flashing aggregative firefly, *Pteroptyx tener. Biological Bulletin* **159** 613–625.

CASE, J.F. (1984) Vision in mating behaviour of fireflies. In: *Insect Communication* (Ed. by Lewis, T.), pp. 195–222. New York: Academic Press.

CASE, J.F. (2004) Flight studies on photic communication by the firefly *Photinus pyralis. Integrative and Comparative Biology* **44** 250–258.

CICERO, J.M. (1983) Lek assembly and flash synchrony in the Arizona firefly *Photinus knulli* Green (Coleoptera: Lampyridae). *Coleopterists Bulletin* **37** 318–342.

COPELAND, J. & MOISEFF, A. (1995) The occurrence of synchrony in the North American firefly *Photinus carolinus* (Coleoptera: Lampyridae). *Journal of Insect Behaviour* **8** 381–394.

COPELAND, J. & MOISEFF, A. (2004) Flash precision at the start of synchrony in *Photuris frontalis. Integrative and Comparative Biology* **44** 259–263.

COPELAND, J., MOISEFF, A. & FAUST, L. (2008) Landing distance in a synchronic North American firefly. *Physiological Entomology* **33** 110–115.

CRATSLEY, C.K. & LEWIS, S.M. (2003) Female preference for male courtship flashes in *Photinus ignitus* fireflies. *Behavioural Ecology* **14** 135–140.

CRATSLEY, C.K. & LEWIS, S.M. (2005) Seasonal variation in mate choice of *Photinus ignitus* fireflies. *Ethology* **111** 89–100.

FAUST, L. (2004) Fireflies as a catalyst for science education. *Integrative and Comparative Biology* **44** 264–265.

FAUST, L. (2010) Natural history and flash repertoire of the synchronous firefly *Photinus carolinus* (Coleoptera: Lampyridae) in the Great Smoky Mountains

National Park. *Florida Entomologist* **93** 208–217.

Fu, X. & Ballantyne, L. (2008) Taxonomy and behaviour of lucioline fireflies (Coleoptera: Lampyridae: Luciolinae) with redefinition and new species of *Pygolu-ciola* Wittmer from mainland China and review of *Luciola* LaPorte. *Zootaxa* **1733** 1–44.

Greenfield, M.D. (2005) Mechanisms and evolution of communal sexual displays in arthropods and anurans. *Advances in the Study of Behaviour* **35** 1–62.

Kaufmann, T. (1965) Ecological and biological studies on the West African firefly *Luciola discicollis* (Coleoptera: Lampyridae). *Annals of the Entomological Society of America* **58** 414–426.

Lewis, S.M. & Cratsley, C.K. (2008) Flash signal evolution, mate choice and predation in fireflies. *Annual Review of Entomology* **53** 293–321.

Lewis, S.M. & Wang, O. (1991) Reproductive ecology of two species of *Photinus* fireflies (Coleoptera: Lampyridae). *Psyche* **98** 293–307.

Lloyd, J.E. (1966) Studies on the flash communication system in *Photinus* fireflies. *University of Michigan Miscellaneous Publications* **130** 1–95.

Lloyd, J.E. (1971) Bioluminescent communication in insects. *Annual Review of Entomology* **16** 97–122.

Lloyd, J.E. (1979) Sexual selection in luminescent beetles. In: *Sexual Selection and Reproductive Competition in Insects* (Ed. by Blum, M.S. & Blum, N.A.), pp. 293–342. New York: Academic Press.

Lloyd, J.E. (1983) Bioluminescence and communication in insects. *Annual Review of Entomology* **28** 131–160.

Mast, S.O. (1912) Behaviour of fireflies (*Photinus pyralis)* with special reference to the problem of orientation. *Journal of Animal Ecology* **2** 256–272.

Maurer, U. (1968) *Some Parameters of Photic Signalling Important to Sexual and Species Recognition in the firefly Photinus pyralis.* M.Sc. Thesis. State University of New York, Stony Brook. 114 pp.

McDermott, F.E. (1964) The taxonomy of the Lampyridae. *Transactions of the American Entomological Society* **90** 1–72.

Michaelidis, C., Demary, K. & Lewis, S.M. (2006) Male courtship signals and female signal assessment in *Photinus greeni* fireflies. *Behavioural Ecology* **17** 329–335.

Moiseff, A. & Copeland, J. (1995) Mechanisms of synchrony in the North American firefly, *Photinus carolinus* (Coleoptera: Lampyridae). *Journal of Insect Behaviour* **8** 395–407.

Moiseff, A. & Copeland, J. (2010) Firefly synchrony: a behavioural strategy to minimize visual clutter. *Science* **329** 181.

Moiseff, A., Copeland, J., Kubke, F. & Faust, L. (2001) Mating behaviour of a synchronous firefly. *Bioluminescence and Chemiluminescence* **11** 153–156.

Ohba, N. (1984) Synchronous flashing in the Japanese firefly, *Luciola cruciata* (Coleoptera: Lampyridae). *Science Report of the Yokosuka City Museum* **32** 23–32.

OHBA, N. (2004) *Mysteries of Fireflies*. Yokosuka City Museum. 199 pp.

OTTE, D & SMILEY, J. (1977) Synchrony in Texas fireflies with a consideration of male interaction models. *Biology of Behaviour* **2** 143–158.

PAPI, F. (1969) Light emission, sex attraction and male flash dialogues in a firefly, *Luciola lusitanica. Monitore Zoologico Italiano* **3** 135–184.

RAU, P. (1932) Rhythmic periodicity and synchronous flashing in the firefly, *Photinus pyralis*, with notes on *Photuris pennsylvanicus. Ecology* **13** 7–11.

RICHMOND, C.A. (1930) Fireflies flashing in unison. *Science* **71** 537–538.

VENCL, F.V. & CARLSON, A.D. (1998) Proximate mechanisms of sexual selection in the firefly *Photinus pyralis* (Coleoptera: Lampyridae). *Journal of Insect Behaviour* **11** 191–207.

VIVIANI, V.R. (2001) Fireflies (Coleoptera: Lampyridae) from southeastern Brazil: habitats, life history, and bioluminescence. *Annals of the Entomological Society of America* **94** 129–145.

WING, S.R. (1985) Prolonged copulation in *Photinus macdermotti* with comparative notes on *Photinus collustrans* (Coleoptera: Lampyridae). *Florida Entomologist* **68** 627–634.

WING, S., LLOYD, J.E. & HONGTRAKUL, T. (1983) Male competition in *Pteroptyx* fireflies: Wing-cover clamps, female anatomy, and mating plugs. *Florida Entomologist* **66** 86–91.

Lampyrid 2012. **2**:99-112

Octopamine and DUM neurons orchestrate the larval firefly aposematic defence

FREDRIC V. VENCL[1,4]*, SHETAL SHAH[2], ADAM GERBER[2] & ALBERT D. CARLSON[3]

[1] *Ecology and Evolution, Stony Brook University, Stony Brook, NY 11794-5245, USA.*
[2] *Ward Melville High School, 380 Old Town Road, East Setauket, NY 11733-3499, USA.*
[3] *Neurobiology and Behavior, Stony Brook University, Stony Brook, NY 11794-5245, USA.*
[4] *The Smithsonian Tropical Research Institute, P. O. Box 2072 Balboa, Ancón, Republic of Panama.*

* *Email: fvencl@life.bio.sunysb.edu*

ABSTRACT Larval fireflies have an aposematic defence. We examined whether the dorsal, unpaired, median (DUM) neurons of the larval central nervous system, which secrete octopamine (OA), play a key role in coordinating enemy detection, the glow warning signal and defensive glandular secretion. We experimentally determined whether 1) lateral abdominal setae signal attack, 2) defensive compounds are stored apart from the haemolymph, and 3) DUM neurons affect glandular secretion. Disturbed larval fireflies injected with methylene blue emitted a clear, bitter substance that was untainted by stain, indicating the presence of reservoirs separate from the haemolymph. Setal manipulation always elicited glows. Glow intensity was dependent on intensity of the stimulus. The latency of response to anterior deflection of the setae was significantly shorter than for posterior deflection. DUM neurons released OA in the vicinity of gland dilator muscles. When perfused in saline containing OA, electrical stimuli elicited dilator tetanus, which was not observed with saline alone. Posterio-lateral setae appear to be phasic, polarised mechanoreceptors signalling threats and inducing both glandular secretion and lantern glow. We hypothesise that when an attack escalates, strong irritation of the lateral abdominal setae induces DUM neurons to flood the haemolymph with OA, which serves to orchestrate the aposematic larval defence response. Octopamine plays an analogous role to noradrenalin in the larval firefly. The similarities point to a very early and perhaps homologous origin of adrenergic molecules in defence. However, support is lacking for the hypothesis that the larval firefly aposematic defence is a 'key innovation.'

Keywords: *Photuris versicolor*, glandular defence, lanterns, fight-or-flight, species swarm

INTRODUCTION

Larvae of the beetle family Lampyridae, or glow-worms, express many attributes associated with aposematic defence. Although they are soft-bodied, slow moving and flightless, they advertise that they are chemically well-defended and unprofitable. They produce a characteristic luminescence from two tiny lantern organs on the penultimate abdominal segment and are unpalatable to many would-be enemies. Lanterns glow spontaneously when larvae crawl, are disturbed, or are attacked by predators (Christensen & Carlson 1982). Many firefly larvae have evolved a strategy that conserves defensive compounds in secretory reservoirs or glands apart from the blood (González et al. 1999; see Fu et al. 2006, 2007). For example, the larvae of *Lampyris noctiluca*, which possess defensive glands along either side of the abdomen (Tyler & Trice 2001), are unacceptable to birds (De Cock & Matthysen 2001; Tyler 2001a), wood ants (Tyler 2001b) and many nocturnal predators (De Cock & Matthysen 2003). Larval luminescence, in conjunction with unpalatability, forms an aposematic anti-predator defence that warns enemies of unprofitability (Underwood et al. 1997; De Cock & Matthysen 1999; Tyler 2002; De Cock & Matthysen 2003). Although the function and neural control of the adult courtship flashes are well known and well studied (reviewed by Lewis & Cratsley 2008), we still know very little about the chemical ecology and evolution of larval firefly aposematic defences.

To escape, a larva must execute a precise sequence of actions involving several different organs that comprise the aposematic defensive strategy. For example, the enemy's attack must be detected, and if attack is imminent, the glow signal must be emitted early enough to minimise the risk of further advances. However, if an attack occurs, the release of glandular exudates that would thwart aggression and reduce the probability of mortality must promptly ensue. Although the neurophysiology of lantern control has been studied in detail, the coordination of responses comprising the complete larval firefly aposematic defence, including glandular secretion, have not been examined.

In the thoraco-abdominal nervous system of insects, dorsal, unpaired, median (DUM) neurons, with their somata located in the midline region, are the major source of the monoamine neurotransmitter octopamine (OA). Previous studies revealed that

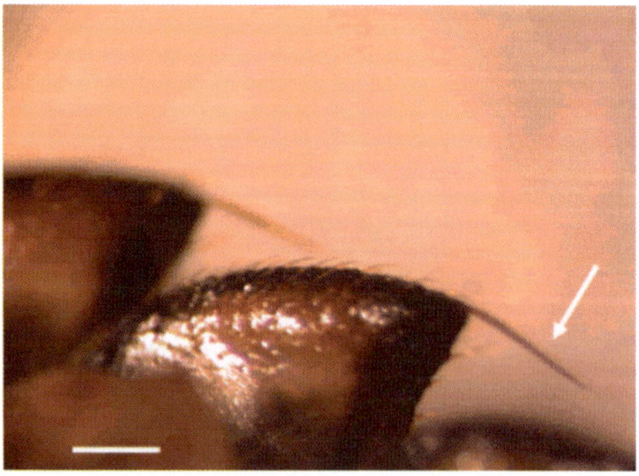

Figure 1 Projecting from the lateral, posterior edge of each abdominal segment is a stiff, long seta (arrow) that serves as mechanoreceptor in *Photuris versicolor* larval fireflies. Scale bar = 500 μm.

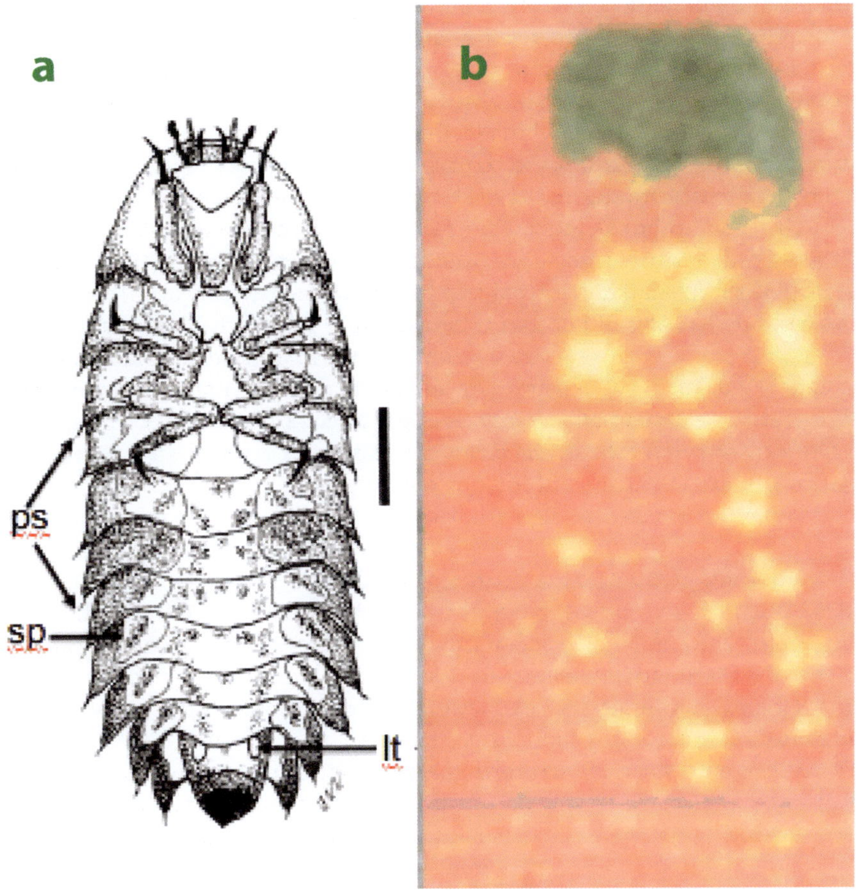

Figure 2 *Photuris versicolor* larva showing the ventral aspect with location of posterio-lateral setae (ps), lanterns (lt) and the lateral sternal plates (sp). **(b)** pH 5 paper stained by *P. versicolor* larva when disturbed: yellow stain adjacent to glandular openings; anterior green stain caused by buccal regurgitate. Scale bar = 1.0 mm.

DUM neurons located in the eighth abdominal ganglion release OA, which activate the larval lanterns of *Photuris versicolor* (Robertson & Carlson 1976; Christensen & Carlson 1982; Christensen et al. 1983; Carlson & Jalenek 1986). Furthermore, each DUM neuron receives both inhibition and excitation from the larval CNS (Carlson 2001). However, the mechanism by which larval fireflies detect enemy propinquity remains unknown. Although firefly larvae lack eyes, they do have an array of stiff, 500 to 600 μm long setae that project from the posterio-lateral edges of the abdominal segments (Figs. 1 & 2a). We postulated that because the setae are strategically situated at the periphery of the larvae, they might signal the immediacy of an enemy attack, and are likely to be the principle but not exclusive mechanoreceptors alerting enemy propinquity. Since the setae project backwards and are deflected toward the body during movement, we further postulated that anterior deflection (away from the larva) would be the primary signal of predator attack.

Our intent was to examine the proximate neural circuitry that coordinates the aposematic defence strategy of *Photuris versicolor* (Fabricius) larvae. We investigated the relationship between the setae and the octopamine-secreting DUM neurons to determine their role in activating the lantern organs. We experimentally addressed three questions: (1) Do larvae possess a separate reservoir system or do they simply release poison into their haemolymph? (2) What role does OA play in glandular emission? (3) What is the neural circuitry linking peripheral mechanoreceptors with the aposematic glow signal and the release of glandular secretions? As a proximate goal, we wished to clarify the role of octopamine in orchestrating the overall aposematic larval defence strategy. Ultimately, we wanted to know if the larval aposematic defence constituted a 'key innovation' that may have enhanced the radiation and diversification of fireflies.

METHODS

Third instar *P. versicolor* larvae were collected in Setauket, NY (40°56'27"N; 73°02'15"W) in September 2005 and 2006 and maintained at 10 °C on moist soil until the following summers, when they moulted to the fourth instar and were large enough to dissect.

To demonstrate the location of the openings of the defensive glands, the ventral surface of a larva was gently pressed against a strip of pH 5 paper, and the location of discharge was photographed. In a second experiment, larvae were injected with methylene blue. If no blue colour appeared where the defensive secretion was released in the previous experiment, we inferred the segregation of the secretion from the haemolymph, perhaps due to presence of reservoirs or glands.

We performed the following experiments to determine the relationship of the posterio-lateral setae to the induction of luminescence. Mechanical manipulation of the setae was carried out to determine whether the setae were involved in signalling predator attack. Phase-shifted larvae (n = 3) were allowed to walk around in a darkened dish and the setae were slowly stroked with a minuten pin attached to a glass rod. The frequencies of luminescent responses were tallied as glows per stroke. In addition, individual setae were deflected forward and backward at different frequencies (1 to 12 cycles per second) to determine if glow intensity was proportional to the intensity of mechanical stimulation. Setal deflection (n = 3) was accomplished by connecting the tip of a seta to a pipette attached to a stimulator-activated speaker. The movement was detected by a force transducer attached to the pipette, and the resulting luminescence (mV) was detected by a photomultiplier aimed at the lantern organs. Both events were recorded on separate channels of a Grass® polygraph. The action potentials in the sensory nerve leading from the posterio-lateral seta to the CNS, elicited by movement of the setae deflected in both forward and backward directions, were recorded by a Grass® P15 preamplifier and the output was transferred to a Hewlett-Packard® 7404A FM tape recorder. The tape recorder's output, along with the output of the force transducer, was displayed on a polygraph at slow speed. To estimate the relationship between deflection rate and luminescence, we used Model I, GLM regression, as implemented by PROC REG (SAS 2004). The latency of setal response, defined as the time (ms) from setal deflection to the first action potential of a burst recorded in the sensory neuron, was measured from the polygraph records at a chart speed of 10 cm/s. Latency measurements were made using a chart speed 15 times faster.

To determine the circuitry connecting the setae, DUM neurons, glands and lanterns, we used a standard larval preparation (see Christensen & Carlson 1982). Larvae (n=3) were severed between the thoracic and abdominal segments and the posterior half was opened by a slit down the midline of the tergite's cuticle. The medial edges were pushed apart and pinned, the gut was removed and the ventral nerve cord and short muscle fibres were exposed. Light micrographs were made of the abdominal region using a Wild® stereomicroscope at a magnification of 50× with a Nikon® FE 35 mm camera. For recording the locations of OA release, a pair of neural connectives were sucked into a stimulating electrode and subjected to a stimulating pulse of one second by a Grass® S44 stimulator. The larval abdomen was perfused with neutral red so that any pinkish stain would indicate OA release sites (Stuart et al. 1974). Finally, dilator muscle contractions were recorded by attaching a Grass® force transducer connected to a polygraph. The nerve connectives were stimulated by alternately perfusing the preparation with firefly saline containing 20 mM OA, or with firefly saline alone as a control (see Christensen & Carlson 1981).

RESULTS

Location of defensive glands

When a larva was gently pressed onto filter paper, a clear substance was emitted from lateral openings between the abdominal and thoracic segments. The secretion turned pH 5 paper yellow, indicating that it was more alkaline than pH 5 (Fig. 2b). Consistent with this finding was the observation that the secretion had a decidedly bitter taste. When we injected methylene blue into the larval haemolymph, none of the blue colour appeared in the secretion.

Using light microscopy, we observed a small group of dilator muscles attached to the dorsal cuticle, lying directly above the intersegmental membrane containing the opening of a sphincter controlling discharge of the glandular secretion (Fig. 3). The dilator muscles were activated by stimulating the ventral nerve cord just anterior to the local ganglion (see below). Reservoirs were located close to the lateral band of

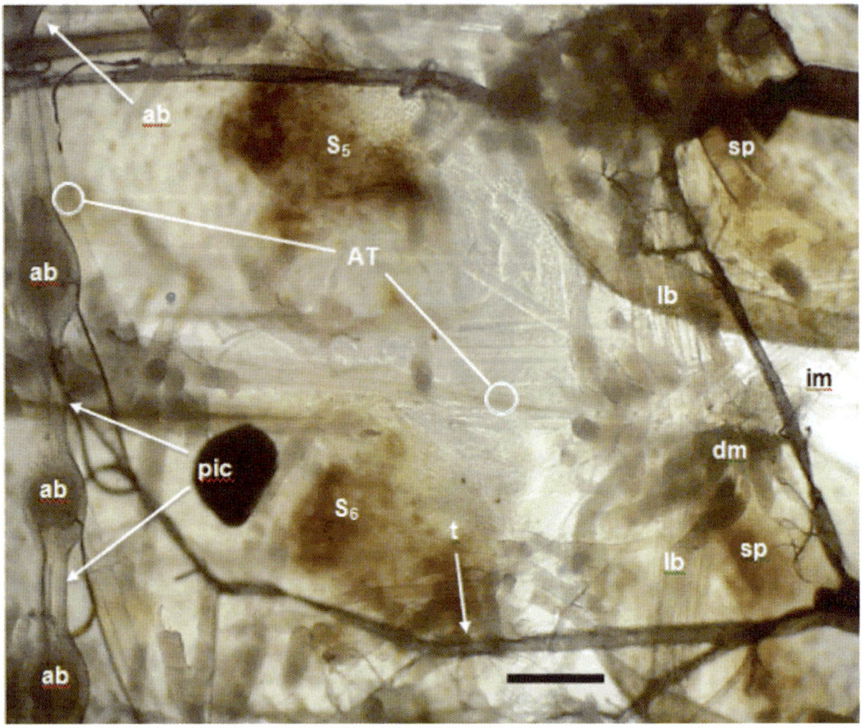

Figure 3 Light micrograph of the dorsal aspect of the area adjacent to the ventral nerve cord in abdominal segments 5 and 6 (**S$_5$** and **S$_6$**) of a *Photuris versicolor* larva, after removal of the dorsal cuticle and digestive tract. The axon trunk (**AT**), originating in the abdominal ganglion (uppermost **ab**), contains axons from the DUM neurons as well as the motor neuron that innervate a group of striate dilator muscle (**dm**) fibres located above the opening of the reservoir in the region of the inter-segmental membrane (**im**). Other abbreviations: **pic**—paired interganglionic connectives of the ventral cord; **sp**—lateral sternal plate (see Fig. 1); **t**—tracheal tubule; **lb**—lateral band of striate muscles responsible for larval curling. Scale bar = 0.10 mm.

three to four muscles that run the length of the larval body on each side. When we perfused the larval abdomen with neutral red, which stains for OA, pinkish areas were observed only in the immediate vicinity of the dilator muscles adjacent to the release sites of the secretion in the inter-segmental membrane.

Activation of warning glow by setae

The mean latency after setal deflection in the forward direction to the onset of action potentials in the afferent neuron was shorter and differed significantly from the mean latency during deflection in the posterior direction (0.69 ± 0.025 and 1.05

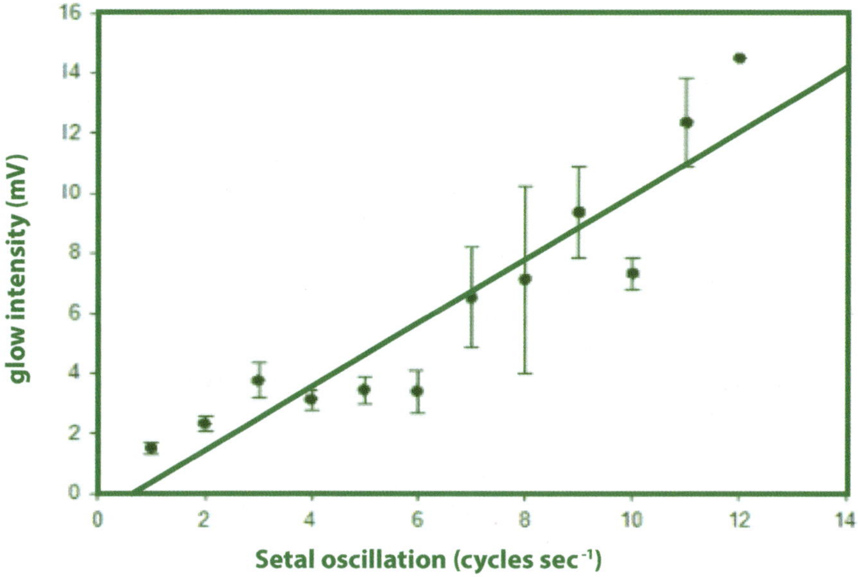

Figure 4 Dependence of lantern luminescent intensity on the rate of setal deflection in *Photuris versicolor* larvae ($y = -0.68 + 1.08x$; $R^2 = 0.85$, $t_{(slope)} = 10.63$, $P < 0.001$), based on pooled responses (n = 195) of 3 larvae.

± 0.037, respectively; $F_{1, 41} = 2.27$, $P < 0.01$). In addition, the variance in latency for anterior deflection was lower than for posterior deflection. No habituation to deflection in either direction was observed. The setal afferents appear not only to directly connect to local DUM neurons in each segment but also, via the ganglionic connective, with the DUM neurons of the 8th segment responsible for driving the lanterns. Gentle movement of the posterio-lateral setae located at the cusp of each abdominal segment elicited luminescent responses 67% of the time (N = 137) in the lanterns. The intensity of lantern luminescence was strongly dependent on the frequency of setal oscillation (Fig. 4).

Octopamine appears to potentiate the activity of the muscles controlling the gland dilators. When recording the contractions of the tiny muscle fibres that likely act to open and close the gland openings, we observed significant differences between the responses of the muscles in saline with and without OA. In firefly saline, single, 1 s shocks to the nerve cord produced occasional single twitches of the muscle fibres (Fig. 5). In firefly saline containing 20 mM OA, a single shock invariably elicited muscle tetanus. Flushing with saline alone returned the muscles to an inactive state.

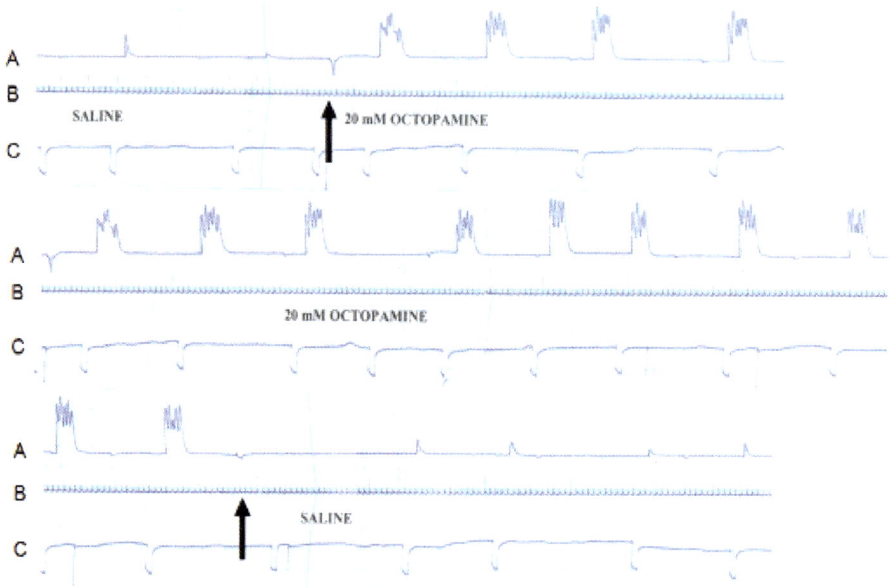

Figure 5 Polygraph traces of the response of larval *Photuris versicolor* dilator muscles to electrical stimulation of the efferent nerve when the abdomen was perfused with saline only, or with 20 mM OA in saline. (A) Contraction response of muscle to stimulation of connective. (B) Time line at one tick per second. (C) Stimulation of neural connectives anterior to the controlling ganglion with single, 2 second electrical pulses. First arrow indicates a switch from perfusing saline alone to saline containing 20 mM OA. Second arrow indicates switch from perfusing saline containing OA back to saline alone.

DISCUSSION

The absence of a blue colour in the defensive secretion of methylene-blue-injected larvae indicated that the secretion is stored in a compartment, or gland, apart from the haemolymph. The presence of a glandular system storing defensive secretions was further supported by two additional findings: the observed presence of dilator muscles, a sphincter and reservoirs in a dissected larva, and the presence of pinkish areas only in the immediate vicinity of the dilator muscles when the larval abdomen was perfused with neutral red, which stains for OA.

The detection of a bitter-tasting secretion emitted from between the abdominal segments when *P. versicolor* larvae are handled is consistent with the finding of the alkaloidal betaine, *N*-methylquinolinium 2-carboxylate, in whole body larval extracts (Gonzáles et al. 1999). We suspect that this substance may be one component in a mixture of deterrent and/or repellent substances secreted from inter-abdominal glands, similar to that found in *Luciola leii* (Fu et al. 2007). Although detailed anatomical analysis will be required to better characterise the morphology of *P. versicolor*, we envision that its glandular structure will resemble that described for *Lampyris, Luciola* or perhaps for some tenebrionid species (Tyler & Trice 2001; Trice et al. 2004; Fu et al. 2007, 2009; see Kendall 1974).

We found strong support for the hypothesis that the posterio-lateral setae, due to their strategic location projecting from the extreme edges of the larva, have two major functions: 1) to alert the larva to imminent predator attack, and 2) to trigger the lanterns to issue the luminescence warning signal via stimulation of the DUM neurons of the eighth abdominal segment. It is significant that mice attacking larval fireflies invariably initiate contact by nibbling on a larva's lateral margins, just where the posterio-lateral setae are situated (D. Tallamy, pers. comm.). The observation that the intensity of lantern glowing was dependent on the intensity of setal deflection implies that a summation of multiple bursts of excitatory action potentials arriving at the abdominal ganglion is required to increase light production. As expected, the posterio-lateral setae appear to be phasic, polarised mechanoreceptors. They respond to anterior deflection with short bursts of action potentials more quickly and with less variation than they do to posterior deflection. Furthermore, they do not habituate. This combination of features permits more rapid and reliable transduction of threat level in the afferent pathway. Not only are these setae strongly tied into the central pathway that leads to excitation of the larval lanterns, but they also activate the DUM neurons serving the glands if stimulation is sufficiently intense. The DUM neurons then release OA onto the two most important targets, the lanterns and glands that form the aposematic defence mechanism.

Neural circuitry of the aposematic defence

Unlike all other motor neurons, which are paired, the DUM neurons are unpaired. Each DUM neuron sends axons to both sides of the body and releases the neuromuscular modulator OA (Hoyle 1975). Our hypothesised neural circuit coordinating glandular secretion and luminescence is illustrated in Figure 6. When a posterio-lateral seta is moved forward (or, to a lesser extent, backward), the sensory nerve is activated and transmits either to a DUM neuron in the eighth abdominal segment, if it is directly anterior to this segment, or to an interneuron in the ganglion, which in turn activates the pathway to the lantern DUM neurons via the connective, initiating luminescence in the paired lanterns.

Octopamine's role as defence orchestrator

Our results support the hypothesis that octopamine, as a neurotransmitter or neurohormone, plays a central role in the coordination of the larval firefly aposematic defence. OA modulates almost every invertebrate organ (Roeder 1999, 2005),

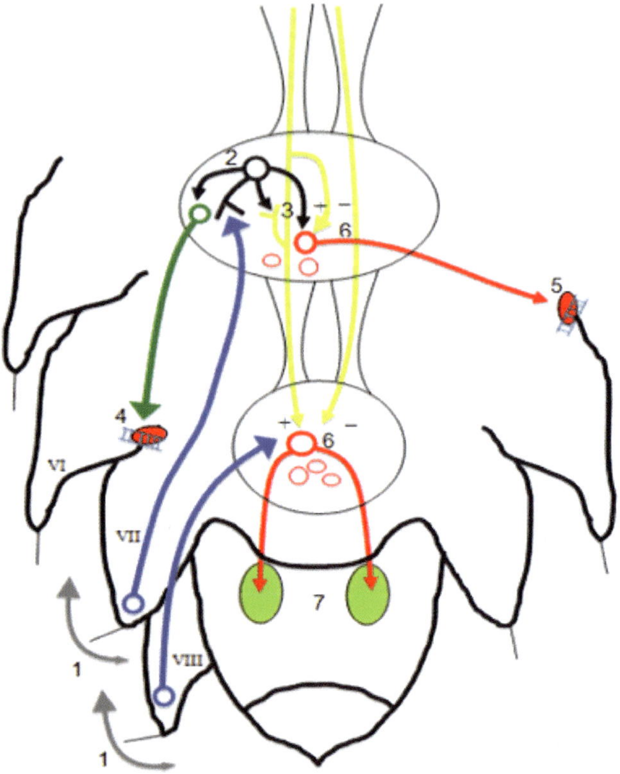

Figure 6 Schematic representation of hypothesised events in the activation of the larval firefly aposematic defence of *P. versicolor*: When the seta of segment VII is deflected (1), the afferent nerve (blue) sends action potentials to an interneuron in the sixth ganglion (2, black circle), which in turn directly activate the inter-ganglionic connective (yellow), the DUM neurons (red circles) in the sixth ganglion (3), as well as the motor neuron (green) controlling inter-segmental glands (4, red ovals) and the release of defensive secretion. If setal stimulation is intense enough, the DUM neurons (6) release octopamine in the vicinity of the inter-segmental glands (5), which causes tetanus in the glandular sphincter muscles. As a second, more direct pathway, deflection of the seta of segment VIII (1) sends action potentials via the sensory afferent (blue), which activates the DUM neurons (6) of the seventh ganglion, driving lantern glowing (7). The inter-ganglionic connective (yellow) integrates inhibitory and excitatory CNS input with information from the setae.

including lantern luminescence in fireflies (Christensen et al. 1983; Carlson & Jalenak 1986). We suspect that in addition to its two known functions as neurohormone and neurotransmitter in invertebrates, OA also acts as a modulator of striate muscle activity at the periphery, chiefly by lowering the threshold of response to the gland-dedicated motor neuron. It appears that our simulated attacks of larvae induced the DUM neurons in the ganglia anterior to those of the eighth segment to flood the haemolymph with OA around the defensive glands, facilitating secretion of defensive chemicals by opening the sphincters of the inter-segmental reservoirs. Our findings are consistent with the role of OA first proposed by Evans and O'Shea (1978) as a modulator of neural inputs to many neuromuscular junctions. Octopamine thus appears to potentiate, *en masse,* the systemic invertebrate 'fight-or-flight' response, much in the same way that noradrenalin does in vertebrates (e.g., Adamo et al. 1995).

Octopamine is exceptional because it is the only nonpeptide neurotransmitter or neurohormone found in invertebrates. Similarities between the octopaminergic and the adrenergic systems are not limited to chemical structure alone. Not only is it structurally related to adrenaline, OA also shares similar *G*-protein receptors (Roeder 2003). Moreover, these molecules appear to play similar physiological roles in the 'fight-or-flight' response to attack. For instance, the insect 'fight-or-flight' response requires coordinated modulation of many organs involved directly and indirectly with defence and escape at different levels. Octopamine is released from neurons onto various peripheral organs, such as larval lanterns and gland dilator muscles, which ultimately results in massive OA pooling in the haemolymph. In addition, OA can trigger the release of adipokinetic hormones from the corpora cardiaca, leading to a liberation of fatty acids from the fat body, which results in higher metabolic rates (heart rate and respiration) and increased fuel supplies to striate muscles by triggering a switch to fatty acid metabolism (reviewed by Roeder 2005). All of these similarities between the adrenergic and octopaminergic signalling systems point to a very early and homologous evolutionary origin.

Can the assembly of the aposematic larval defence suite be reconstructed?

The complexity of defence—compound biosynthetic pathways, morphology and patterns of deployment—all suggest adaptation and multi-trait defence suite formation, as for complex structures in general. Two key steps were necessary for suite assembly: glandular compartmentalisation of costly and/or toxic repellent or deterrent substances, and co-option of the neural motor circuitry for coordinated lantern and glandular control activated by the setal sensory afferents.

From a macroevolutionary perspective, recent phylogenetic studies and character reconstructions suggest that larval firefly luminescence is ancestral and conserved within the radiation of the Lampyridae, and appears to have evolved after chemical unpalatability but before adult courtship signals (Branham & Wenzel 2003; Stanger-Hall et al. 2007; Jeng 2008). Moreover, natural selection not only fostered the evolution of a warning signal in order to conserve deterrent and repellent chemicals, which are likely far more costly to produce than light, but also the neural control mechanisms to precisely coordinate the glow signal with glandular secretion, to affect escape *in extremis*. We suspect that firefly diversification was promoted, at least in part, by the evolution of an effective and efficient aposematic defence in larvae, the firefly's most vulnerable developmental stage.

Is the lampyrid larval aposematic display a key innovation?

Biological traits that promote lineage diversification via mechanisms that increase the rate of speciation and/or decrease the rate of extinction are regarded as 'key innovations.' Key innovation hypotheses are most frequently studied by means of correlations between the degree to which multiple origins of a trait are associated with clades of unusual size (e.g., species 'swarms' or 'flocks'). Phylogenetic evidence that a key innovation increases the diversification rate of a lineage is most commonly drawn from diversity comparisons between sister clades (thus, with equal evolutionary age) or from the distribution of the number of lineages (branch lengths) through time. However, neither the diversities nor the relative lengths of terminal branches suggest that the aposematic larval firefly defence suite covaries with diversification in lampyrids when compared to sister-group cantharid taxa (see Sagegami-Oba et al. 2007). Diversity analyses are limited because the evolution of aposematism is not well understood and the classification of the families within the cantharidoids (cantharids, lampyrids, lycids, etc.) remains controversial. Another approach might entail comparing divergence rates between the aposematic traits and housekeeping traits. If rapid evolution in the clade bearing the aposematic suite could be demonstrated compared to divergence rates in housekeeping traits such as morphology, size, and trophic ecology, then an advantage could be inferred. This approach could provide insights into processes underlying species radiations, but to date, such comparisons have been largely qualitative. At present, strong support is lacking for the hypothesis that the larval firefly aposematic defence represents a "key innovation".

ACKNOWLEDGEMENTS

The authors are grateful to K. Studholme for providing the photomicrographs. This project involved students who participated in the Research Program for High School Students, sponsored by the Center for Science, Mathematics and Technology Education at Stony Brook University. This is contribution #1204 from the Department of Ecology and Evolution at Stony Brook University.

REFERENCES

Adamo, S.A., Linn, C.E. & Hoy, R.R. (1995) The role of neurohormonal octopamine during "fight or flight" behavior in the field cricket *Gryllus bimaculatus*. *Journal of Experimental Biology* **198** 1691–1700.

Branham, M.A. & Wenzel, J.W. (2003) The origin of photic behavior and the evolution of sexual communication in fireflies (Coleoptera: Lampyridae). *Cladistics* **19** 1–22.

Carlson, A.D. (2001) Central control of larval luminescence. In: *Progress and Current Applications. Proceedings of the 11th International Symposium on Bioluminescence and Chemiluminescence, Asilomar Conference Grounds, Pacific Grove, California, USA, 6-10 September 2000* (Ed. by Case, J.F., Herring, P.J., Robison, B.H., Haddock, S.H.D., Kricka, L.J. & Stanley P.E.), pp. 149–152. Singapore: World

Scientific Publishing Co. Pte. Ltd.

Carlson, A.D. & Jalenek, M. (1986) Release of octopamine from the photomotor neurons of the larval firefly lanterns. *Journal of Experimental Biology* **122** 453–457.

Christensen, T.A. & Carlson, A.D. (1981) Symmetrically organized dorsal unpaired median DUM neurones and flash control in the male firefly, *Photuris versicolor. Journal of Experimental Biology* **93** 133–147.

Christensen, T.A. & Carlson, A.D. (1982) The neurophysiology of larval firefly luminescence: Direct activation through four bifurcating (DUM) neurons. *Journal of Comparative Physiology* **148** 503–514.

Christensen, T.A., Sherman, T.G., McCaman, R.E. & Carlson, A.D. (1983) Presence of octopamine in firefly photomotor neurons. *Neuroscience* **9** 183–189.

De Cock, R. & Matthysen, E. (1999) Aposematism and bioluminescence: experimental evidence from glow-worm larvae (Coleoptera: Lampyridae). *Evolutionary Ecology* **13** 619–638.

De Cock, R. & Matthysen, E. (2001) Do glow-worm larvae use warning coloration? *Ethology* **107** 1019–1033.

De Cock, R. & Matthysen, E. (2003) Glow-worm larvae bioluminescence operates as an aposematic signal upon toads (*Bufo bufo*). *Behavioral Ecology* **14** 103–108.

Evans, P. D. & O'Shea, M. (1978) An octopaminergic neuron modulates neuromuscular transmission in the locust. *Nature* **270** 257–259.

Fu, X.H., Meyer-Rochow, V.B., Tyler, T., Suzuki, H. & De Cock, R. (2009) Structure and function of the eversible organs of several genera of the larval firefly (Coleoptera: Lampyridae). *Chemoecology* **19** 155–168.

Fu, X.H., Ohba N., Vencl, F.V. & Lei, C. (2006) Life cycle and behaviour of the aquatic firefly, *Luciola leii* (Coleoptera: Lampyridae), from Mainland China. *Canadian Entomologist* **138** 860–870.

Fu, X.H., Vencl, F.V., Ohba, N., Meyer-Rochow, V.B., Lei, C. & Zhang, Z. (2007) Structure and function of the larval eversible glands of the aquatic firefly *Luciola leii* (Coleoptera: Lampyridae). *Chemoecology* **17** 117–124.

González, A., Schroeder, F., Meinwald, J. & Eisner, T. (1999) N-methylquinolinium 2-carboxylate, a defensive betaine from *Photuris versicolor* fireflies. *Journal of Natural Products* **62** 378–380.

Hoyle, G. (1975) Evidence that insect dorsal unpaired median (DUM) neurons are octopaminergic. *Neurobiology* **9** 43–57.

Jeng, M.L. (2008) *Comprehensive Phylogenetics, Systematics, and Evolution of Neoteny of Lampyridae (Insecta: Coleoptera)*. Ph.D. Thesis. University of Kansas.

Kendall, D.A. (1974) The structure of defence glands in some Tenebrionidae and Nilioidae (Coleoptera). *Transactions of the Royal Entomological Society of London* **125** 437–487.

Lewis, S.M. & Cratsley, C.K. (2008) Flash signal evolution, mate choice and predation in fireflies. *Annual Review of Entomology* **53** 293–321.

Robertson, H.A. & Carlson, A.D. (1976) Octopamine: Presence in firefly lantern

suggests a transmitter role. *Journal of Experimental Zoology* **195** 159–164.
ROEDER, T. (1999) Octopamine in invertebrates. *Progress in Neurobiology* **59** 1–31.
ROEDER, T. (2003) Metabotropic histamine receptors: nothing for invertebrates? *European Journal of Pharmacology* **466** 85–90.
ROEDER, T. (2005) Tyramine and octopamine: ruling behavior and metabolism. *Annual Review of Entomology* **50** 447–477.
SAGEGAMI-OBA, R., TAKAHASHI, N. & OBA, Y. (2007) The evolutionary process of bioluminescence and aposematism in cantharoid beetles (Coleoptera: Elateroidea) inferred by the analysis of 18S ribosomal DNA. *Gene* **400** 104–113.
SAS. (2004) Vesion 9.0. The SAS Institute, Cary, North Carolina, USA.
STANGER-HALL, K., LLOYD, J.E. & HILLIS, D.M. (2007) Phylogeny of North American fireflies (Coleoptera: Lampyridae): Implications for the evolution of light signals. *Molecular Phylogenetics and Evolution* **45** 33–49.
STUART, A.E., HUDSPETH, A.J. & HALL, Z.W. (1974) Vital staining of specific monoamine-containing cells in the leech nervous system. *Cell Tissue Research* **153** 55–61.
TRICE, E., TYLER, J. & DAY, J.C. (2004) Description of pleural defensive organs in three species of firefly larvae (Coleoptera, Lampyridae). *Zootaxa* **768** 1–11.
TYLER, J. (2001a) A previously undescribed defense mechanism in the larval glow-worm *Lampyris noctiluca*? *Coleopterist* **10** 38.
TYLER, J. (2001b) Are glow-worms *Lampyris noctiluca* (Linnaeus) (Lampyridae) distasteful? *Coleopterist* **9** 148.
Tyler, J. (2002) *The Glow-worm*. Sevenoaks: Lakeside Printing Ltd. 75 pp.
TYLER, J. & TRICE, E. (2001) A description of a possible defensive organ in the larva of the European glow-worm *Lampyris noctiluca*. *Coleopterist* **10** 75–78.
UNDERWOOD, T.J., TALLAMY, D.W. & PESEK, J.D. (1997) Bioluminescence in firefly larvae: a test of the aposematic display hypothesis. *Journal of Insect Behavior* **10** 365–370.

Lampyrid 2012. **2**:113-122

Survey of species diversity, distribution and abundance of fireflies in Singapore

Su Hooi Chan[1*], Lesley A. Ballantyne[2] & Colleen A.L. Goh[3]

[1] *Sungei Buloh Wetland Reserve, National Parks Board, 301 Neo Tiew Crescent, Singapore 718925, Republic of Singapore.*
[2] *School of Agricultural and Wine Sciences, Charles Sturt University, PO Box 588, Wagga Wagga 2678, Australia.*
[3] *31 Hindhede Walk, #03-12, Singapore 587967, Republic of Singapore.*
**Email: chan_su_hooi@nparks.gov.sg*

ABSTRACT Surveys were carried out from February 2009 to April 2010 in Singapore to document the species diversity, distribution and abundance of fireflies at selected sites. Five different habitats have been surveyed. The survey has revealed the possible occurrence of seven genera and 11 species (several unidentified). Three genera are from the subfamily Luciolinae, two from the Lampyrinae, one from the Ototretinae and one from the family Rhagophthalmidae. We have confirmed the original records of *Pteroptyx valida* Olivier, *Stenocladius* sp. and *Diplocladon* sp. but not of *Pteroptyx bearni* Olivier and *Lucidina wallacei* Pic. The survey is establishing which areas have a higher species abundance and diversity. Some insight into firefly habitat preference is emerging. Results of the survey will allow identification of some of the key firefly habitats in Singapore, and a better understanding of the habitat requirements of the fireflies. This information will guide future efforts in the conservation of fireflies in Singapore.

Keywords: Firefly, diversity, distribution, habitat, Singapore

INTRODUCTION

Singapore is the second smallest country in Asia and a tropical island city-state with a land area of 710 sq km. It consists of one main island and about 60 smaller off-shore islands. Within this city-state are 22 nature sites, comprising four legally protected nature reserves covering 33.26 sq km (about 4 % of Singapore's total land area) and 18 nature areas that support natural ecosystems recognised for their significant biodiversity. The climate is characterised by uniform temperatures between 23–34 °C, with an average daily humidity of 84 %, and an average annual rainfall of 2,300 mm (National Parks Board 2010).

Among the more attractive insects found in Singapore, butterflies and dragonflies have been well studied. Fireflies are iconic and attractive insects but little is known about the firefly fauna of Singapore. As fireflies are affected by pollution (Nada et al. 2009) and may therefore be good indicators of the health of natural ecosystems, it is all the more important to fill the gap in knowledge of fireflies. The firefly fauna in Singapore has not previously been well surveyed and studied. Based on available literature, five species in four genera were originally recorded from Singapore. The five species are *Pteroptyx valida* Olivier (Ohba 1993; Ohba & Sim 1994), *Pteroptyx bearni* Olivier (McDermott 1966), *Lucidina wallacei* Pic (McDermott 1966), *Stenocladius* sp. (Ohba et al. 1996) and *Diplocladon hasseltii* (Haneda 1985). Existing information pertaining to the occurence of firefly species and their distribution in Singapore is scant. It was therefore proposed that a survey be carried out to determine and document the species diversity, distribution and abundance of fireflies in Singapore. A Firefly Study Group consisting of National Parks Board staff and volunteers was formed to carry out the surveys. The preliminary results of the survey, conducted between February 2009 and April 2010 are presented.

METHODS

Survey sites

A total of fourteen sites were selected from nature reserves, nature areas and natural green areas located on the main island and two smaller nearby islands.

Five different habitats were represented: scrubland/grassland, mangrove, freshwater swamp, primary forest and secondary forest (Fig. 1). The key mangrove sites that are easily accessible and with existing trails were selected. The Central Nature Reserve (CNR) in central Singapore was chosen as it contains one of the oldest lowland dipterocarp forests and a hill dipterocarp forest comprising primary and secondary forests, and a freshwater swamp forest. The scrubland in north-west Singapore was selected because it is a locally known firefly site. Pulau Semakau, an offshore island located south of Singapore, was chosen because it is among the key mangrove sites in Singapore while Pulau Ubin, located northeast of Singapore, is one of the important nature areas in Singapore.

Survey protocol

One to three sites were visited monthly from February 2009 to April 2010 (except for March and September 2009 and February and March 2010). Of the fourteen sites, eight were visited once, five were visited twice and one site had three visits. Surveys were conducted from 7.30–10.30 pm. Two groups, each with five volunteers, covered different areas within each site. The number of fireflies, as indicated by their flashes, was recorded from walks along existing trails at the forest fringe or within the forests. The time and GPS location for each sighting was also recorded. Adult flight habits, larval habits, and light patterns of both adult and larva, were recorded. Soil moisture was taken using a moisture meter, with readings ranging

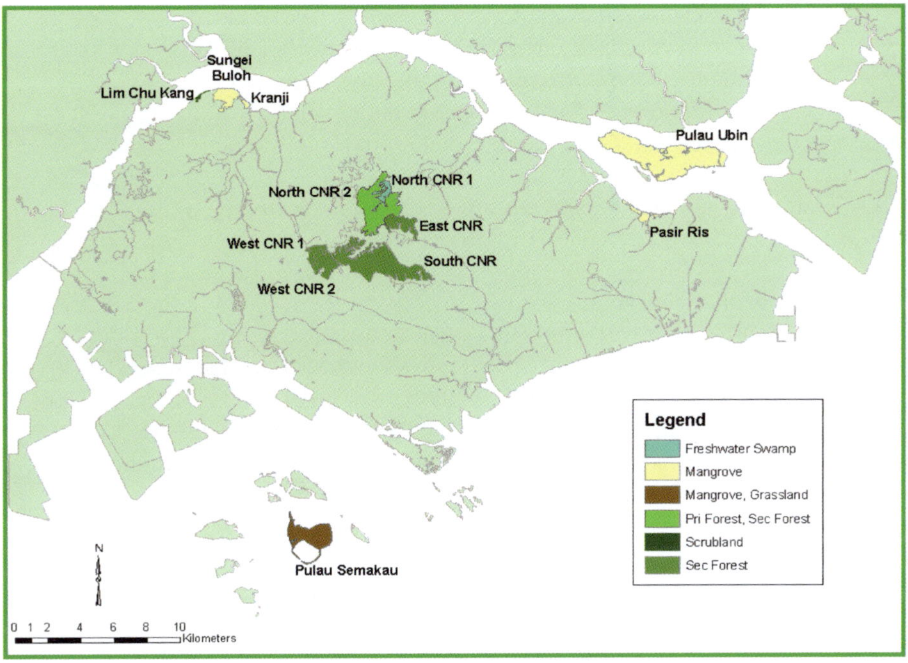

Figure 1 Locations of the 12 sites surveyed for fireflies in Singapore, showing habitat types. Sites may have habitat types other than those surveyed in this study.

from 1 to 3 categorised as dry, 4 to 7 as moist and 8 to 10 as wet. Vegetation density was recorded as dense or sparse. Specimens were collected during the surveys using hand-held nets or collection vials. Photographs of the specimens were taken *ex-situ* and sent to experts for identification. Attempts were made to rear larvae collected. Larvae were provided with snails collected from the same site, and larval feeding choices and behaviour were noted. Voucher specimens of each species are deposited at the Raffles Museum of Biodiversity Research, Singapore.

RESULTS

The survey revealed the occurrence of seven genera and eleven species, three of which were identified as *Pteroptyx valida*, *P. malaccae* Gorham and *Colophotia praeusta* Eschscholtz, the latter two being new records for Singapore (Table 1). Abundances at each site are shown in Table 2.

Habitat and Site	Firefly species [1]
Mangrove	
Sungei Buloh	*Pteroptyx valida* Olivier
Kranji	*P. valida*
Pasir Ris	*P. valida*
Pulau Ubin	*P. valida* ?
Pulau Semakau Site 1	*P. valida, Pteroptyx malaccae* Gorham
Primary Forest	
North CNR2 Site 1	*Luciola* sp. 1
Secondary Forest	
North CNR2 Site 2	*Luciola* sp. 1, *Pyrocoelia ?fumigata* Gorham
East CNR	*Pyrocoelia ?fumigata*
South CNR	?*Luciola* sp. 1, Lucioline sp. 1, *Pyrocoelia ?fumigata*
West CNR1	Lucioline sp. 2, *Pyrocoelia ?fumigata, Stenocladius* sp., *Diplocladon ?hasseltii* Gorham
West CNR2	?*Luciola* sp. 1, *Pyrocoelia ?fumigata, Stenocladius* sp., *Diplocladon ?hasseltii*
Freshwater Swamp	
North CNR1	*Luciola* sp. 2, Lucioline sp. 1, *Pyrocoelia ?fumigata, Diaphanes ?subapicalis/dorsalis/raapi*
Scrubland/Grassland	
Pulau Semakau Site 2	*P. valida*
Lim Chu Kang	*Colophotia praeusta* Eschscholtz

[1] In the Lampyridae: *P. valida, P. malaccae, C. praeusta, Luciola* sp. 1, *Luciola* sp. 2, Lucioline sp. 1, Lucioline sp. 2 (Luciolinae); *Pyrocoelia ?fumigata, Diaphanes ?subapicalis/dorsalis/raapi* (Lampyrinae); and *Stenocladius* sp. (Ototretinae). In the Rhagophthalmidae: *Diplocladon ?hasseltii. Stenocladius* sp. and *Diplocladon ?hasseltii* identified from larval forms. Unidentified Lucioline larva #1 was collected in West CNR1 and 2, and Lucioline larva #2 in West CNR1 (not listed in table).

Table 1 Occurrence of firefly species at fourteen selected sites in Singapore.

Representatives of four new genera were recorded, *Colophotia praeusta,* and three tentatively identified as *Pyrocoelia ?fumigata* Gorham, *Luciola* sp. 1 and sp. 2, and *Diaphanes ?subapicalis/dorsalis/raapi*. Two unidentified Lucioline species were also collected, and two morphospecies of Lucioline larvae. *Pteroptyx valida* was found at all mangrove sites as the sole species, except at Pulau Semakau where *P. malaccae* was also collected. Outside the mangrove areas, species collected from single sites were *C. praeusta* (Lim Chu Kang scrubland), *Diaphanes ?subapicalis/dorsalis/raapi, Luciola* sp. 2 (both North CNR1 freshwater swamp), and Lucioline sp. 2 (West CNR1 secondary forest), while the other species, i.e., *Pyrocoelia ?fumigata, Diplocladon ?hasseltii* Gorham, *Stenocladius* sp., *Luciola* sp. 1, and Lucioline sp. 1 were found in at least two CNR sites. *Pyrocoelia ?fumigata* was recorded

Site	Date	Adult numbers	Larval numbers
Lim Chu Kang	17 Apr 2009	Approx. 130	-
Sungei Buloh	27 Feb 2009	5	12
Kranji	6 Nov 2009	5	14
Pasir Ris	3 Apr 2009	Approx. 40	Approx. 30
Pasir Ris	18 Dec 2009	Approx. 70	Approx. 30
Pulau Ubin	3 Jul 2009	2	-
Pulau Semakau Site 1 & 2	28 Apr 2010	Approx. 60	-
North CNR1	31 Jul 2009	9	-
North CNR1	22 Jan 2010	17	1
North CNR2 Site 1 & 2	31 Jul 2009	9	-
North CNR2 Site 2	2 Oct 2009	13	3
North CNR2 Site 1 & 2	22 Jan 2010	9	5
East CNR	19 Jun 2009	2	-
East CNR	31 Jul 2009	3	-
South CNR	29 May 2009	17	8
West CNR1	8 May 2009	10	26
West CNR1	28 Aug 2009	24	11
West CNR2	8 May 2009	4	8

Table 2 Numbers of fireflies found in each site.

from all the CNR sites. Adult and larval behaviour and site conditions in which they were found are described below.

Pteroptyx valida (Luciolinae) was the dominant species in all the mangrove sites and was not found in other habitats except the grassland site 500 m from the mangrove site at Pulau Semakau. Soil moisture in these sites ranged from 4 to >10. A few larvae were found in areas with soil moisture reading as low as 2, and a few on the edge of the granite footpath. Most of the larvae were seen among damp leaf litter, comprising primarily *Hibiscus tiliaceus* leaves, and on the soil in the mangrove and mangrove fringes where *Acrostichum aureum* and *Acanthus ebracteatus* were present. Two larvae were seen on tree trunks approximately 20 cm above ground. The adults were observed flashing non-synchronously on *Avicennia alba* and *Panicum maximum*, flying and flashing among *Rhizophora apiculata* and *Sonneratia alba,* flying from ground level to high up in tall trees, and flashing on soil. Attempts made to rear the larvae collected were successful, and the larvae were observed feeding on *Subulina octona*. Larvae were observed pulling snails to a preferred location, i.e., under leaf litter, to feed.

Pteroptyx malaccae (Luciolinae) was recorded only on Pulau Semakau, in a densely vegetated mangrove where the soil moisture reading exceeded 10. About 10 individual adults that were likely *P. malaccae* were observed flashing synchronously in *Sonneratia alba* (ca. 10 m height), and some were seen flying and flashing at 2.0–2.5 m among vegetation and at the mangrove fringes where *Rhizophora apiculata* and *Sonneratia alba* were present.

Colophotia praeusta (Luciolinae) was found only in scrubland in Lim Chu Kang in an area with few trees and with thick grass and soil moisture readings from 4.0 to 7.5. It was seen flashing and flying low among and above grasses and weeds up to about 1 m in height, and perching and flashing on weeds. In this area, at about 1930 hrs, fireflies began flying out of the grass and weeds upwards, and small congregating groups of about 30 individuals were observed flashing non-synchronously in three trees (ca. 7 m in height), one of which was identified as *Acacia auriculiformis*. That these fireflies were also *Colophotia* could not be confirmed as the area in which they occurred was not accessible for sampling.

Luciola sp. 1 (Luciolinae) is an undetermined *Luciola* sp. from the *praeusta* complex. It was found in both primary and secondary forests at North CNR2, South CNR and West CNR2 in areas with dense vegetation and lower soil moisture (1 to 4). Individuals were observed emitting a continuous glow when in flight, but flashed upon landing on vegetation. It frequented open areas, flying at the forest fringe at less than 3 m from the ground. It was often seen glowing or flashing on plants at about 0.5 m from the ground. A mating pair was observed on a plant at about 10 cm from the ground, and the female was seen pulling the male behind her when disturbed. On a few occasions, individuals resting on plants were observed to drop from the leaf onto the ground and "play dead" when disturbed, a self-protective behaviour.

Luciola sp. 2 (Luciolinae) is not consistent with any described species and is probably new. It was found only at North CNR1 in the freshwater swamp in an area with dense vegetation, damp leaf litter and higher soil moisture (4 to >10). The adults were seen congregating within a very small area at a specific location and the majority were flashing on a species of fern identified as *Blechnum finlaysonianum*. Some individuals were observed flying low and flashing among the ferns and at the fringe of the forest vegetation. All were observed below 2 m.

Lucioline sp. 1 (Luciolinae) was found as a single pair in the freshwater swamp at North CNR1, flashing on a leaf at about 1 m above ground. A few individuals of this morphospecies were also found in the secondary forest of South CNR. They were observed flying and flashing at not more than 3 m above ground, or stationary and flashing on plants. We found it in areas with dense vegetation and near or at freshwater streams.

Lucioline sp. 2 (Luciolinae) was found only at West CNR1 in the secondary forest, in areas with dense vegetation and moist soil (7). We observed a single female flashing on soil and on the root tip of an uprooted tree at about 1 m from the ground. Only a few sightings were made and only the females of this species have been recorded. We have not observed it in flight thus far.

Two Lucioline larval forms, named Lucioline Larva #1 and #2, respectively, were also collected but not successfully reared to adult and remain unidentified. A few Lucioline Larva #1, which resembled larvae of *Pteroptyx valida*, were collected from the secondary forest of West CNR1 and West CNR2 with dense vegetation and low soil moisture readings ranging from 2 to 5. However, leaf litter was observed to be damp in the areas it frequented. Individuals on soil and leaf litter were seen producing continuous or intermittent glows with long intervals. In captivity, it was observed feeding on three species of snails, *Subulina octona*, *Bradybaena similaris* and *Lamellaxis clavulinus*. It was seen attacking and pulling the snail to a corner on the soil before feeding on it. It often burrowed into the soil to hide. Two Lucioline Larva #2 were found, and only at West CNR1 in the secondary forest with dense vegetation and soil moisture reading of 4. The larvae, which were collected from leaf litter near a stream, were observed to glow intermittently. In captivity, they clung to the underside of dried leaves, and they fed on *Subulina octona*.

Pyrocoelia ?fumigata (Lampyrinae) appeared to be the most common species in the secondary forest. It was recorded at all the secondary forest sites and the freshwater swamp within the CNR in central Singapore, but not at the primary forest site in North CNR2. Soil moisture readings in those areas in which it occurred ranged from 1.5 to 8.0. The larvae produced intermittent glows on leaf litter and on soil at the forest fringe or further into the vegetation. Solitary adult males were often seen at the forest fringe in open areas flying from about 0.3 m to 3.0 m emitting continuous light. Larvae in captivity fed on six species of snails of which five were identified: *Hemiplecta humphreysiana*, *Bradybaena similaris*, *Subulina octona*, *Quantula striata* and *Amphidromus atricallosus perakensis*. On one occasion, a larva was seen attacking a snail many times bigger than itself along a tarmac road at North CNR2 during the day. Both the larva and snail were collected, and the larva was observed making several attempts to feed on the snail, but eventually giving up, possibly because the snail was very large. On two other occasions, a larva was observed removing the faecal matter of a snail before feeding on it. Another larva was observed climbing on top of the shell of a snail and attacking it from above. The snail protected itself from the attack by moving its shell left and right continuously and by secreting a fluid forming into foam that rapidly hardened and protected the snail.

Diaphanes ?subapicalis/dorsalis/raapi (Lampyrinae) was found only in the freshwater swamp at North CNR1. A single male adult was found in a moist area that had dense vegetation, and was observed emitting continuous light and flying upwards to a height of about 3 m.

Stenocladius sp. (Ototretinae) larvae were found in the secondary forest of West CNR1 and West CNR2 in areas with soil moisture readings of 3. The larvae emitted continuous glows on soil and crawled on leaf litter at the fringe of dense forest vegetation. They were very mobile and glowed briefly when disturbed. This species appeared to have a preference for areas with less pristine forest. In captivity, the larva was often seen exposed on soil or hiding under a dried leaf. *Stenocladius* larvae appeared to be more active than *Diplocladon* larvae.

Diplocladon ?hasseltii (Rhagophthalmidae) larvae were recorded at West CNR1 and West CNR2. This species shares a similar habitat with the *Stenocladius* sp., and the larvae were found at the fringe of dense secondary forest vegetation. It was recorded in both low and high soil-moisture areas, with moisture readings ranging from 1.5 to 8.0. Larvae were observed on soil as well as on granite chips, emitting a continuous glow. When disturbed, the larvae curled up or moved quickly to hide in the soil or under leaf litter. In captivity, the larvae spent most of the time hiding in a chamber in the soil or hiding under a dried leaf on the soil.

DISCUSSION

Of the 11 species of fireflies found in the present surveys, six are recorded for the first time, i.e. *Pteroptyx malaccae*, *Colophotia praeusta*, *Pyrocoelia ?fumigata*, *Luciola* sp. 1 and sp. 2, and *Diaphanes ?subapicalis/dorsalis/raapi*. The survey has confirmed the original records of *Pteroptyx valida* Olivier (Ohba 1993; Ohba & Sim 1994), *Stenocladius* sp. (Ohba et al. 1996) and *Diplocladon* sp. (Haneda 1985) but not of *Pteroptyx bearni* Olivier (McDermott 1966; Ballantyne & McLean 1970) and *Lucidina wallacei* Pic (McDermott 1966). For *Pteroptyx bearni*, the possibility that the original locality was incorrect should be considered. *Lucidina wallacei* was not found during the surveys, possibly because it is a diurnal species.

It appears that all habitats are important for conservation as different habitats had different characteristic species. Species richness appears higher in the CNR in central Singapore compared to elsewhere on the island while firefly abundance appears highest at the Lim Chu Kang, Pasir Ris and Pulau Semakau sites. Within the CNR, North CNR1, West CNR1 and West CNR2 appear to support more species. The greater variety of habitats in the CNR is likely the reason for the higher species richness there. The freshwater swamp at North CNR1 and the survey sites consisting of less pristine secondary forest at West CNR1 and CNR2 support higher numbers of species.

With the exception of Pulau Semakau mangroves, which had both *P. valida* and *P. malaccae*, the mangrove sites generally supported only one species of firefly, *P. valida*, and in abundance. Likewise, the Lim Chu Kang scrubland yielded large numbers of the sole species recorded for that site, *C. praeusta*. These three firefly species may be habitat-specific, with the former two limited to mangroves and the latter confined to scrubland. The collection of *P. malaccae* from only the relatively inaccessible and pristine mangroves at Pulau Semakau and not from other mangrove sites indicates that this species may be sensitive to habitat disturbance.

The tolerance of different firefly species to soil moisture levels appeared to vary, with *Luciola* sp. 1, *Stenocladius* sp. and the two Lucioline larval forms apparently able to survive drier conditions. *Pyrocoelia* sp. and *Diplocladon* sp. were found in both wet and dry areas, while *P. valida*, *P. malaccae*, *C. praeusta*, *Luciola* sp. 2, *Diaphanes* sp., Lucioline sp. 1 and Lucioline sp. 2 appeared to prefer wetter areas.

More sampling is needed to determine whether there is any habitat-specificity in *Luciola* sp. 2 and *Diaphanes* sp., which were found only in the freshwater swamp during the two site visits. While Lucioline sp. 2 and Lucioline Larva #2 share a similar habitat with other species found in the secondary forest, they were found only in one locality. It is possible that the two species are dependant on the habitat conditions that the survey site at West CNR1 provides.

If an area supported only one species of firefly, greater numbers were seen and these fireflies tended to congregate. Areas supporting several species of fireflies contained smaller numbers of fireflies and solitary individuals were usually observed.

Additional sampling in these sites may yield more species and should produce useful data on the seasonal abundance and habitat requirements of the fireflies of Singapore.

ACKNOWLEDGEMENTS

We wish to thank Dr Ming-Luen Jeng, Dr Nobuyoshi Ohba and Sonny Wong for their contributions to the identification of the fireflies. We thank Rachel Lim for preparing the maps and for assistance in spatial analysis. We are also indebted and grateful to the following colleagues and friends for providing kind assistance in the field, sharing their expertise and giving constant support and encouragement: Wong Tuan Wah, James Gan, Ng Sock Ling, Leong Tzi Ming, Kelvin Lim, Lua Hui Kheng, Tan Siong Kiat, Yeo Suay Hwee, Lim-Ho Chee Len, Chew Ping Ting, Yang Shufen, Lim Wei Ling, Robin Ngiam, Adeline Chia, John Low, Tay Soon Lian, How Choon Beng, Wong Kum Sang, Linda Goh, Halilah Ahmad and Firefly Study Group members. SHC also thanks her family for support and encouragement throughout the study.

REFERENCES

Ballantyne, L.A. & McLean, M.R. (1970) Revisional studies on the firefly genus *Pteroptyx* Olivier (Coleoptera: Lampyridae: Luciolinae: Luciolini). *Transactions of the American Entomological Society* **96** 223-305.

Haneda, Y. (1985) *Luminous Organisms*. Tokyo: Kouseisha Kouseikaku. 318 pp.

McDermott, F.A. (1966) Lampyridae. In: *Coleopterorum Catalogus Supplementa, Pars 9* (Ed. by Steel, W.O.), Vol. 9, 149 pp. Gravenhage: W. Junk.

Nada, B., Kirton, L.G., Norma-Rashid, Y. & Khoo, V. (2009) Conservation efforts for the synchronous fireflies of the Selangor River in Malaysia. In: *Diversity and Conservation of Fireflies. Proceedings of the International Symposium on*

Diversity and Conservation of Fireflies, Queen Sirikit Botanic Garden, Chiang Mai, Thailand, 26-30 August 2008 (Ed. by Napompeth, B.), pp. 160-171. Chiang Mai: Queen Sirikit Botanic Garden.

NATIONAL PARKS BOARD. (2010) *Fourth National Report to the Convention on Biological Diversity*. Singapore. 67 pp.

OHBA, N. (1993) Luminescence insect investigation of Singapore and Malaysia. *Annual Report of the Yokosuka City Museum*. 6 pp.

OHBA, N., GOTO, Y. & KAWASHIMA, I. (1996) External morphology, color-marking patterns and habits of the larval stage in genus *Stenocladius* (Coleoptera: Lampyridae). *Science Report of the Yokosuka City Museum* **44** 21-31.

OHBA, N. & SIM, S.H. (1994) The morphology, behaviour and life cycle of *Pteroptyx valida* (Coleoptera: Lampyridae) in Singapore. *Science Report of the Yokosuka City Museum* **42** 1-11.

Lampyrid 2012. **2**:123-126

The distribution and abundance of fireflies in natural and agricultural areas of Pakistan

WASEEM AKRAM*, MUHAMMAD SAJID TABASSUM & KHURAM ZIA

Department of Agri-Entomology, University of Agriculture, Faisalabad, Pakistan.
**Email: areeba14@yahoo.com*

ABSTRACT Over 70% of the population of Pakistan is involved in agriculture. Cultivated areas have shown an increase in pesticide consumption over the past few years. Some issues of concern are resistance development in pests and the death of non-target insects, which include natural enemies and insects of aesthetic value such as fireflies. Fireflies are well known for their beautiful night light displays and are indicators of a clean environment. The present study was carried out in semi-natural ecosystems, and agro-ecosystems where the common use of pesticide sprays has polluted the water sources. Estimates of firefly abundance indicate that sprayed areas had fewer fireflies than unsprayed areas. Estimates of firefly abundance in semi mountainous areas indicated that the vegetation close to streams and water channels where aquatic life was abundant have high populations of fireflies.

Keywords: Fireflies, distribution, agro-ecosystems, natural ecosystems, pesticide pollution, Pakistan

INTRODUCTION

Fireflies are indicators of a clean environment and their presence indicates a healthy ecosystem. The extensive use of pesticides in agricultural areas in Pakistan may have drastically reduced firefly populations, which in the past were abundant. Very little is known about the effects of fertilisers and pesticides on firefly habitats. There have been no efforts to preserve these creatures because no work relating to their identification and distribution has ever been carried out in Pakistan. The present study was carried out in Pakistan to find out for the first time their distribution in relation to selected land use types, i.e., agricultural areas and forests, and to record their seasonal abundance.

METHODS

The present study was carried out from March 2009 to June 2010 at four sites in Faisalabad (local sites, < 24 km from Faisalabad) and four sites in Rawalpindi (outstation sites, 335-385 km from Faisalabad). These sites have one or more existing firefly colonies but vary in pollution level (including agrochemical runoff) and land uses. Surveys were conducted over three nights (from sunset to midnight) every month, to quantify the abundance of fireflies. Firefly abundance for each colony was estimated through visual counts of the average number of flashes per minute, obtained for five replicates. The length and height of the display area was also recorded. Firefly populations were also sampled for laboratory rearing and observation, and for wet and dry preservation (ten and five individuals, respectively, per site). Specimens of adults and immatures are deposited at the University of Faisalabad. The following were recorded for each site: elevation, ground temperature (using an infrared temperature gun), relative humidity (RH, in %, using a hygrometer), presence of pollution, and land use categories present, i.e., agricultural land under cultivation and fallow agricultural land, grassland, forest and urban areas. Cultivated land is sprayed while fallow land is not.

Local sites: Water channels around the University of Agriculture, Gutwala Forest, Jahranwala road and Ghulam Muhammad Abad were visited. Gutwala Forest is a recreational area with trees, bushes, tall grasses and a water canal. Surveys in Ghulam Muhamad Abad were conducted in agricultural and non-agricultural areas, where the clogged water channels are heavily polluted due to garbage dump leachate and sewerage water. Jahranwala road, historically an agricultural area, has over the past few years seen an increase in residential development, being close to Faisalabad.

Outstation sites: Rawalpindi sites were: Ayub Park, a semi-natural forest with artificial lakes, plantation, forest, parks and amusement spots; the historical Wah Garden, which comprises a stream and pond in a garden landscaped with bushes, tall grasses and trees; Rawal Lake View Park, a forest park next to Rawal Lake, which has faced critically low water levels for the past eight years due to drought.

RESULTS

At (local) Faisalabad sites, fireflies began appearing in early April with peak numbers recorded in April, thereafter declining to almost zero by the end of May. Mean temperatures in April and May were 26 and 30°C, respectively, increasing to a high of 45°C in June. Relative humidity (RH) dropped from April to June, which recorded RH of under 40%. Fireflies were seen between 7.00 to 10.30 pm. Maximum activity in April and May was from 7.15 to 9.15 pm and 8.15 to 9.15 pm, respectively (Table 1).

At the (outstation) Rawalpindi sites, fireflies were seen from April to mid-June, and from mid-July to September, with peaks in April and October.

Fireflies were most abundant at Ayub Park and Wah Garden, which recorded over 100 flashes per minute. These two sites also maintained high levels of firefly activity even in the months in which firefly activity was declining at other sites.

Site	Location	Elevation (m asl)	Site characteristics [1]							Firefly activity[2] (flashes/ minute)
			W	L	Land use categories present					
					C	N	G	F	U	
Local sites										
Univ. Agric. Faisalabad	31°25′00″N 73°04′59″E	175	P, T	•	•	•	•		•	88
Ghulam Muhamad Abad	31°25′00″N 73°04′59″E	170	P, T	•	•	•				56
Jahranwala road	31°20′00″N 73°26′00″E	184	P, T	•	•	•				81
Gutwala Forest	31°25′00″N 73°04′59″E	169	P			•	•	•		69
Outstation sites										
Ayub Park	33°36′00″N 73°04′01″E	484	P			•	•	•		112
Wah Garden	30°14′10″N 72°44′10″E	133	P			•	•	•		121
Rawal Lake	33°42′00″N 73°07′00″E	1800	P			•	•	•		61

[1] W, Water source (P, Permanent; T, Temporary); L, Pollution; C, Cultivated (sprayed); N, Fallow land (not sprayed); G, Grassland; F, Forest; U, Urban.

[2] Presents peak activity for the respective sites, from records taken between March 2009 and June 2010. The average number of flashes per minute was obtained for five replicates.

Table 1 Sites surveyed for fireflies in Pakistan. Local and outstation sites were up to 24 km and 335 - 385 km, respectively, from Faisalabad.

The Faisalabad University and Jahranwala Road recorded a moderate range (80-89) of flashes per minute. Two local sites, i.e., Ghulam Muhamad Abad and Gutwala Forest, and one outstation site, i.e., Rawal Lake, recorded firefly activity in the lower range (50-69 flashes per minute).

DISCUSSION

Both sites that recorded the most firefly activity were located away from Faisalabad city, and in forested, non-agricultural areas with minimal pollution. Rawal Lake, which had similar site characteristics, however, recorded low firefly activity, perhaps due to extremely low water levels there. Also possessing similar site characteristics but recording low firefly activity was Gutwala Forest, a local site. The factors contributing to this result are not known.

A common characteristic of sites that were polluted was the presence of agricultural activity, where heavy pesticide applications are the norm. The lowest recorded firefly activity, which was at Ghulam Muhamad Abad, was also observably more polluted than the other sites that were considered polluted, i.e., Faisalabad University and Jahranwala road. These observations are consistent with previous findings in Japan that link declining firefly populations to pesticide applications (Tabaru et al. 1970).

The decline in firefly numbers from the end of April to June at most sites appeared related to rising temperatures and increasing dryness at the sites. Higher levels of firefly activity maintained at Wah Garden and Ayub Park during those months could be due greater shade in those areas, which thus remained cooler with constant water levels.

This preliminary assessment of areas where fireflies are abundant and areas where firefly populations may be threatened, and the identification of factors that may be contributing to their decline, are a positive first step towards firefly conservation efforts in Pakistan.

REFERENCES

Tabaru, Y., Kouketsu, T., Oba, M. & Okafuji, S. (1970) Effects of some organophosphonis insecticides against the larvae of Genji firefly, *Luciola cruciata* Motschulsky (Lampyridae: Coleoptera) and their prey, Japanese melanial snail, *Semisulcospira bensoni* Phillipi (Pleuroceridae: Mesogastropoda). *Japanese Journal of Sanitary Zoology* **21** 178-181 (In Japanese).

Lampyrid 2012. **2**:127-137

Insect species co-existing with the Papua New Guinea firefly *Pteroptyx effulgens* share aspects of appearance and behaviour

Nobuyoshi Ohba[1] & Victor Benno Meyer-Rochow[2]*

[1] *Yokosuka City Museum, Yokosuka 238-0016, Japan.*
[2] *Jacobs University Bremen, Faculty of Engineering & Science, D-28759 Bremen, Germany; and Department of Biology, University of Oulu, SF-90014 Oulu, Finland.*
**Email: b.meyer-rochow@jacobs-university.de*

ABSTRACT Male individuals of the New Guinean firefly *Pteroptyx effulgens* start emitting flashes of light from particularly large trees after sunset. Individual flash cycles with approximately 0.8 second intervals gradually synchronise as it gets darker, so that ultimately the entire tree appears to be sending out flashing signals in synchrony. Daytime collections showed that, on trees used by *P. effulgens* for their luminescent displays, a variety of insects live that resemble the firefly in morphology and colouration of the pronotum and wings. This article examines the kinds of mimicry that could account for the assemblage of similar looking insects in one and the same location, which number at least 20 species belonging to 11 families and four orders. The significance of this phenomenon lies in the fact that a clearly defined space (i.e., the tree) is occupied by a population of one dominant species (i.e., *P. effulgens*) allowing the evolution and co-existence of numerous similar-looking insects.

Keywords: Lampyridae, Lycidae, mimicry, predator avoidance, co-existence, Batesian, Müllerian, Mertensian, defence, toxic

INTRODUCTION

If two or more animal species resemble each other or share certain behavioural or physiological characteristics, we may be dealing with a phenomenon of mimicry (Edmunds 1974; Wickler 1990). For instance, should a predatory species resemble or copy the behaviour of a harmless one in order to gain access to a food source, we speak of "aggressive mimicry." A well known example of this kind of mimicry is when female *Photuris* fireflies mimic the mating signals of *Photinus* females. *Photinus* males that approach are preyed on by the mimic. In another form of mimicry (known as Mertensian mimicry), highly toxic prey species as well as totally harmless ones may mimic a moderately poisonous model.

Cases where only one or more harmless species mimic a moderately distasteful, toxic, or hard-to-handle species are classified as Batesian mimicry. For Batesian mimicry to suceed, the mimic species must be less common than the model, from which unconditioned predators may learn avoidance. Finally, if several noxious species share the same warning colouration or behavioural patterns, the species complex is said to employ Müllerian mimicry. If the predator forms one avoidance image for all prey species with that shared appearance and/or behaviour, this can work to the advantage of all participant prey species.

In situations such as the one we describe below, in which trees are inhabited by large populations of the firefly *Pteroptyx effulgens* and a variety of insects resembling *P. effulgens* as well as each other, the type (or types) of mimicry involved are not immediately apparent. Fireflies have long been known to be distasteful to most predators, e.g., birds, lizards, frogs, fish (Sexton 1960; Lloyd 1973), spiders and centipedes (Fu et al. 2007). It would, thus, seem that fireflies would be the model mimicked by other species. However, as we shall describe below, several of the insect species resembling *P. effulgens*, especially those belonging to the family Lycidae, are themselves toxic. Characterising mimic-model relationships is further complicated by species similar to *P. effulgens* in colouration and body shape but not body size, which are active when the nocturnal *P. effulgens* are not.

P. effulgens males gather on one tree over several days and start emitting flashes of blue-green light soon after sunset (Ohba & Shimoyama 2009). The flash cycle of individual fireflies gradually synchronises, with regular ca. 0.8 sec intervals between flashes. From a distance, the whole tree appears to be sending out regular flashes of blue-green light into the darkness of the tropical night. On the same tree, several look-alike insect species, including the considerably less widespread *Pteroptyx tarsalis,* co-exist with *P. effulgens* (Ohba & Shimoyama 2009). With the exception of *P. tarsalis,* the rest of the look-alike species are non-luminescent, resembling fireflies in general body shape and colour pattern of pronotum and wings, but not necessarily size. Some of these similar-looking insects have been seen mating on the same leaf on which *P. effulgens* was found, but distinguishing model and mimic and determining whether this can be ascribed to mimicry is not straightforward. What is certain is that most of the insects resembling *P. effulgens* do not spend their larval periods on the same tree as the firefly, but use the trees only as adults for resting and mating.

Since it has long been known that many firefly species are distasteful to predators and, when attacked, can reflexively 'bleed' (Blum & Sannasi 1974; Ohba & Hidaka 2002; Fu et al. 2007) haemolymph containing lucibufagins (Eisner et al. 1997), the glow and body colouration of adults (Underwood et al. 1997) and larvae are often interpreted as aposematic signals (De Cock & Matthysen 1999, 2001, 2003; De Cock 2009). If the production and secretion of toxic defence chemicals is a universal feature of all firefly species, it would be tempting to predict that the very numerous *P. effulgens* serve as a model for other species that would mimic the distasteful species to gain survival value. However, fireflies of the genus *Photuris*, which are incapable of producing distasteful chemicals themselves, sequester lucibufagins by feeding on *Photinus* species (Eisner et al. 1997). The assemblage of numerous insect species co-occurring with *P. effulgens* and resembling one another raises complex questions about who mimics whom and why. In the following we shall try to shed some light on this problem.

METHODS

Study sites

Populations of *P. effulgens* and similar-looking insect species, occuring on the same tree at Open Bay (New Britain: NB) and Kaw (New Ireland: NI) in Papua New Guinea (Fig. 1A & B), were studied. *Pteroptyx effulgens* is a firefly of 7 mm total body length, with an orange-yellow pronotum and black elytra (Fig. 1C). Individuals

Figure 1 **A & B.** Display trees (indicated by arrow, species unknown) of *Pteroptyx effulgens* in: A) New Britain, and B) New Ireland (Papua New Guinea). **C.** The firefly *Pteroptyx effulgens*. **D.** Example of one analysed photograph (cf. Table 2, photo no. 3) with 8 *P. effulgens* and 3 mimics visible.

Taxon[1]	New Britain No. species	No. individuals	New Ireland No. species	No. individuals
Pteroptyx effulgens	1	249	1	376
Mimics				
COLEOPTERA	**16**	**43**	**8**	**131**
Lycidae	11	38	3	120
Cantharidae	1	1	2	8
Cleridae	1	1		
Chrysomelidae	1	1	1	1
Carabidae	1	1		
Oedemeridae	1	1		
Harpalidae			1	1
Staphylinidae			1	1
LEPIDOPTERA	**1**	**1**		
ORTHOPTERA			**1**	**1**
DERMAPTERA (Forficulidae)			**1**	**1**
TOTAL	**17**	**293**	**11**	**509**

[1]Order subtotals in bold.

Table 1 Numbers of *Pteroptyx effulgens* and similar-looking insects collected from display trees in two Papua New Guinea locations.

can be seen on some highly visible trees around the villages. The abundant forests and groves around these display trees provide suitable conditions for the larvae to complete their development. The fireflies appear not to be restricted to a particular species of tree, but rather select display trees based on size, shape, and foliage. Large trees with abundant leafage are preferred gathering places. The firefly aggregations we studied ranged from a few thousand to several tens of thousands of individuals. The tree in NB was about 40 m tall, while the one in NI measured approximately 7 m in height and represented a different (unidentified) species.

Sampling method

Individuals of *P. effulgens* and similar-looking insect species were observed for two weeks in August 1991 in NB and two weeks in October 2003 in NI. The trees on which they occurred were observed and photographed. During the study period, *P. effulgens* and similar-looking insect species were randomly collected with a 45 cm wide sweep net on a 3 m long handle. Ten sweeps at a time between 13.00 and 16.00 h were carried out. Collections from the 40 m tall tree were made by local helpers, but insects on the lower branches and leaves up to a height of 3 m were not disturbed. Because the highest density of insects (by visible inspection) occurred at about 1.5 m above the ground, it was at this level that our photographs were taken. *Pteroptyx effulgens* was identified to species while the remaining specimens were sorted to morphospecies and the number of morphospecies and individuals

	PHOTO 1		PHOTO 2		PHOTO 3		PHOTO 4	
	Species	N	Species	N	Species	N	Species	N
Pteroptyx effulgens	1	20	1	16	1	8	1	10
Mimics								
COLEOPTERA								
Lycidae	2	3	1	2	2	2	2	2
Cantharidae	1	2	1	1	1	1	-	-
TOTAL	4	25	3	19	4	11	3	12

Table 2 Numbers of *Pteropytx effulgens* and similar-looking insects counted in four photographs taken of *P. effulgens* display trees in New Britain, Papua New Guinea. Numbers of species (Species) and number of individuals (N) for each photograph are shown side by side.

were recorded for each insect order. For the order Coleoptera, sorting and recording was further done to family level. Many dozens of close-up photographs of *P. effulgens* and similar-looking insect species were taken at various viewing angles to document assemblages of different but similar-looking species at close proximity to one another. The photographs were the basis for analysing and describing the morphological features of the different species encountered.

Analyses also made use of four photographs from NB on which *P. effulgens* and similar-looking insects were found together and could clearly be identified. The photographs were carefully examined in order to count and distinguish the numbers of individual *P. effulgens* and similar-looking insects and to identify the mimics preferably to family level. The percentage of insect individuals resembling *P. effulgens* was then calculated for the various photographs. A sample photograph is included in Figure 1D.

RESULTS

Of the 802 individuals of *P. effulgens* and insects resembling *P. effulgens* collected from NB and NI, 22.1% comprised insects resembling *P. effulgens* with 20 species in 11 families and 4 orders (Table 1). Coleoptera made up 85% of species resembling *P. effulgens* and 12 of the 20 species were lycids. In terms of abundance, 98.3% (n = 177) of all insects resembling *P. effulgens* were beetles and 90.8% (n = 174) of these beetles were lycids. Apart from Lycidae, the beetle families Chrysomelidae, Cantharidae, Cleridae, Staphylinidae, Oedemeridae, and Harpalidae, and the insect orders Lepidoptera, Dermaptera, and Orthoptera were represented for *effulgens*-like species (Fig. 2A).

The percentage of *effulgens*-like individuals was higher for NI than for NB (26.1%, n = 509, and 14.8%, n = 293, respectively). Also more *effulgens*-like species were found in NI than NB (n = 17 and n = 10, respectively). This difference was largely attributed to more *effulgens*-like lycid species found in NI than NB (n = 11 and n = 3, respectively).

From the four photographs analysed with regard to the NB population, a total of 54 *P. effulgens* individuals were counted together with 13 individuals of similar-looking insects (3 species in 2 coleopteran families, Table 2). Both lycids and cantharid species resembling *P. effulgens* were captured in three photos while the fourth photo did not record cantharids. The percentage of *effulgens*-like individuals recorded ranged from 15.8 to 27.3% (Table 2). Some of the insects resembling *P. effulgens* were seen to mate in bright daylight, visible to might-be predators, on the same leaf on which *P. effulgens* was resting (Fig. 2B).

DISCUSSION

Assemblages of insects resembling each other

Several possible explanations can be offered for the profusion of species co-occurring with *P. effulgens* on the same tree. If species co-evolved similar features, there has to be an evolutionary explanation, and convergence could be one such explanation as shown by Sagegami-Oba et al. (2007). In order to demonstrate that similar colour markings in cantharid, lampyrid and lycid beetles represent an ancestral trait already present in proto-beetles prior to the division of these beetle families within the Cantharoidea, genetic studies similar to those of Day (2009) and Oba (2009) on the evolution of mechanisms of beetle bioluminescence would be required. Yet, even if the assumed proto-beetles shared certain features, the question remains why these features were conserved in the different family lineages.

Since some of the insects resembling *P. effulgens* were seen to mate in bright daylight, visible to would be predators, on the same leaf on which *P. effulgens* were resting, it would seem as if *P. effulgens* as well as the species resembling *P. effulgens* are avoided by predators. During the day, bioluminescent flashing cannot function as an aposematic signal and therefore body colouration and shape needs to adequately convey the message of distastefulness to predators, though for diurnally-active lampyrid larvae this explanation may not apply (cf. De Cock 2009). Body size may be less important than colouration and shape, because most scavenging birds and lizards detect their prey monocularly with either their right or left eye, which makes immediate distance estimation difficult (Hockey 1997).

If *P. effulgens* represented an acceptable prey for a predator, it would be hard to explain how this species can form such huge aggregations throughout the year without being attacked by predators. Although we are not aware of chemical analyses to demonstrate toxic substances in *P. effulgens*, we can assume that its predators avoid it for some reason, most likely distastefulness. But the reason/s other insects resemble *P. effulgens* are less straight forward. The majority of the insects resembling *P. effulgens* in this study were lycid beetles and it is well known that this family contains highly toxic species (Crowson 1981), which are models for mimics from other beetle families, e.g., Buprestidae, Elateridae, Cantharidae, Lampyridae, Cleridae, Meloidae, Oedemeridae, Pyrochroidae (Moore & Brown 1989).

Figure 2 A. Insect species resembling *Pteroptyx effulgens* in New Britain and New Ireland in Papua New Guinea. From the families: a-f, Lycidae; g, Staphylinidae; h – i, Cantharidae; j, Carabidae (Harpalinae); k, Chrysomelidae (Coleoptera). From the orders: l, Lepidoptera; m, Orthoptera. **B.** Individuals of a lycid species copulating on a leaf on which *Pteroptyx effulgens* (the latter not seen in this photograph) rests.

In the USA, lycids are known to serve as models for mimics from the coleopteran families Chrysomelidae, Cantharidae and Cerambycidae, and from the orders Diptera (e.g., *Rhamphomyia* sp., *Plecia* sp. and *Arge* sp.) and Lepidoptera (e.g., *Harrisina* sp. and *Cisseps fulvicolis*). While mimics from the Orthoptera are not mentioned, the female orthopteran possessing pronotum and abdomen colouration of *P. effulgens* and mingling with other *effulgens*-like insects in this study could well be a Batesian mimic. The effect is further enhanced by having green legs that blend with the colour of the leaf on which the nymphal orthopteran rests. In the adult, wings cover its abdomen and the entire body assumes a green colouration for camouflage. Therefore, this species of Orthopteran appears to employ two different defence strategies during its lifetime.

Comparisons between species composition and abundance

In NB the 44 insects resembling *P. effulgens* represented two orders, seven families and 17 species, and made up 14.8% of the total number of collected insects (Table 1). Most commonly encountered were *effulgens*-like Coleoptera, which represented 94.1% of the total number of *effulgens*-like insects and belonged to 16 species from 6 families, of which the Lycidae, with 68.8% of all *effulgens*-like beetles, were the dominant family. In NI, too, Lycidae were dominant among 6 families and eight species of Coleoptera collected, representing 90.9% (120) of the 136 *effulgens*-like insects collected. Altogether, amongst 376 *P. effulgens*, insects resembling the latter represented 10 species from seven families and three orders.

Considered together, insects resembling *P. effulgens* at the two sites studied in NB and NI contained representatives of 4 orders, 11 families and 20 species (Fig. 3). In NB a greater variety of species was seen, but NI contained a larger number of orders and families. In both locations, species belonging to Lycidae were the most abundant of the *effulgens*-like insects. The ratios of *effulgens*-like insects to *P. effulgens* were 44 : 249 and 133 : 376 for NB and NI, respectively.

What kind of mimicry are we dealing with?

One of the principles of Batesian mimicry is that the mimetic species must be less common than the model. Given that *P. effulgens* outnumbered by far the other species resembling it, the other species could be Batesian mimics. For representatives of the Orthoptera, Dermaptera and some beetles of the family Cleridae this is most likely the case as these groups are not known to be distasteful or poisonous. The large number of lycid species, many of which could be highly toxic (Dettner 1987; Lewis 2009) could indicate Müllerian chemical mimicry (Dettner & Liepert 1994), where a complex of similar-looking, noxious species enable predators to develop an avoidance image.

However, another scenario is possible and represents perhaps the most likely explanation for the assemblage of insect species resembling each other at the two locations studied by us. Assuming *P. effulgens* is moderately toxic or distasteful and

Figure 3 A collection of mimics obtained from the study including 17 species of Coleoptera (Lycidae, Chrysomelidae, Cantharidae, Cleridae, Carabidae (Harpalinae)) and small moth (Lepidoptera). Other species were collected but are not shown.

predators that try to feed on *P. effulgens* feel sick afterwards, but do not die, then some species can evolve that are potentially lethal to a predator and some that are harmless, but both resembling the moderately poisonous model. Both the highly toxic species and the harmless ones need to occur in smaller numbers than the model, for if the deadly species was too abundant and an animal preying upon it would die, no learning and, thus, no future avoidance could occur.

In Mertensian mimicry, potentially deadly species (in this case some lycids) as well as harmless species would mimic a moderately poisonous model from which predators could learn avoidance. The main requirement is that the model, firstly, needs to be most abundant and, secondly, restricted to one location. Both conditions apply to *P. effulgens* (as well as to *P. tener*, although perhaps not everywhere, but according to Ohba and Wong, 2004, at least in Selangor), but not to the Indonesian synchronising firefly *Pyrophanes appendiculata* E. Olivier. This latter firefly, although abundant, may gather in one tree, but will then suddenly fly off to a different tree. Potential mimics cannot follow, and it would seem that for this reason insects mimicking *P. appendiculata* have never been found despite extensive searches by one of us (N. Ohba). This research demonstrates that the study of mimicking insects has to be based not just on comparisons of abundance and on morphological and physiological features of the species involved. It also needs to consider the wider ecological and ethological questions of habitat preference and behaviour.

REFERENCES

Blum, M.S. & Sannasi, A. (1974) Reflex bleeding in the lampyrid *Photinus pyralis*: Defensive function. *Journal of Insect Physiology* **20** 451-460.

Crowson, R.A. (1981) *The Biology of the Coleoptera.* New York and London: Academic Press.

Day, J. (2009) Beetle bioluminescence: a genetic and enzymatic research review. In: *Bioluminescence in Focus - A Collection of Illuminating Essays* (Ed. by Meyer-Rochow, V.B.), pp. 255-276. Trivandrum, India: Research Signpost.

De Cock, R. (2009) Biology and behaviour of European lampyrids. In: *Bioluminescence in Focus - A Collection of Illuminating Essays* (Ed. by Meyer-Rochow, V.B.), pp. 161-200. Trivandrum, India: Research Signpost.

De Cock, R. & Matthysen, E. (1999) Aposematism and bioluminescence: experimental evidence from glow-worm larvae (Coleoptera; Lampyridae). *Evolutionary Ecology* **13** 619-639.

De Cock, R. & Matthysen, E. (2001) Do glow-worm larvae (Coleoptera; Lampyridae) use warning coloration? *Ethology* **107** 1019-1033.

De Cock, R. & Matthysen, E. 2003. Glow-worm larvae bioluminescence (Coleoptera: Lampyridae) operates as an aposematic signal upon toads (*Bufo bufo*). *Behavioural Ecology* **14** 103-108.

Dettner, K. (1987) Chemosystematics and evolution of beetle chemical defences. *Annual Review of Entomology* **32** 17-48.

Dettner, K. & Liepert, C. (1994) Chemical mimicry and camouflage. *Annual Review of Entomology* **39** 129-154.

Edmunds, M. (1974) *Defence in Animals.* Harlow: Longman.

Eisner, T., Goetz, M.A., Hill, D.E., Smedley, S.R. & Meinwald, J. (1997) Firefly "femmes fatales" acquire defensive steroids (lucibufagins) from their firefly prey. *Proceedings of the National Academy of Sciences of the USA* **94** 9723-9728.

Fu, X., Vencl, F.V., Ohba, N., Meyer-Rochow, V.B., Lei, C. & Zhang, Z. (2007) Structure and function of the eversible glands of the aquatic firefly *Luciola leii* (Coleoptera: Lampyridae). *Chemoecology* **17** 117-124.

Hockey, P.A.R. (1997) Eyes in the sky: Understanding avian vision. *Africa Birds and Birding* **2** 50-53.

Lewis, S.M. (2009) Bioluminescence and sexual signalling in fireflies. In: *Bioluminescence in Focus - A Collection of Illuminating Essays* (Ed. by Meyer-Rochow, V.B.), pp. 147-159. Trivandrum, India: Research Signpost.

Lloyd, J.E. (1973) Firefly parasites and predators. *Coleopterists Bulletin* **27** 91-106.

Moore, B.P. & Brown, W.V. (1989). Graded levels of chemical defence in mimics of lycid beetles of the genus *Metriorrhynchus* (Coleoptera). *Journal of the Australian Entomological Society* **28** 229-233.

Oba, Y. (2009) On the origin of beetle bioluminescence. In: *Bioluminescence in Focus - A Collection of Illuminating Essays* (Ed. by Meyer-Rochow, V.B.), pp.

277-290. Trivandrum, India: Research Signpost.

Ohba, N. & Hidaka, T. (2002) Reflex bleeding of fireflies and prey-predator relationship. *Scientific Reports of the Yokosuka City Museum* **49** 1-12 (in Japanese with English abstract).

Ohba, N. & Shimoyama, A. (2009) The synchronous flashing signal of *Pteroptyx effulgens* in Papua New Guinea is used by *P. tarsalis* to form aggregations. In: *Bioluminescence in Focus - A Collection of Illuminating Essays* (Ed. by Meyer-Rochow, V.B.), pp. 229-242. Trivandrum, India: Research Signpost.

Ohba, N. & Wong, C.H. (2004) External morphology and ecological study of the firefly, *Pteroptyx tener* at Kampong Kuantan, Selangor, Malaysia. *Scientific Reports of the Yokosuka City Museum* **51** 1-33.

Sagegami-Oba, R., Takahashi, N. & Oba, Y. (2007) The evolutionary process of bioluminescence and aposematism in cantharoid beetles (Coleoptera: Elateroidea) inferred by the analysis of 18S ribosomal DNA. *Gene* **400** 104-113.

Sexton, O.J. (1960) Experimental studies of artificial Batesian mimics. *Behaviour* **15** 245-252.

Underwood, T.J., Tallamy, D.W. & Pesk, J.D. (1997) Bioluminescence in firefly larvae: a test of the aposematic display hypothesis (Coleoptera: Lampyridae). *Journal of Insect Behavior* **10** 365-370.

Wickler, W. (1990) *Mimikry*. Frankfurt: Fischer.

Lampyrid 2012. **2**:138-141

Regional diversity of fireflies of the subfamily Luciolinae (Coleoptera: Lampyridae) in Sri Lanka

W.M.C.D. Wijekoon, H.C.E. Wegiriya* & C.N.L. Bogahawatta

Department of Zoology, University of Ruhuna, Matara, Sri Lanka.
**Email: hemantha@zoo.ruh.ac.lk*

ABSTRACT According to the documented records of the National Museum of Sri Lanka, 63 species of fireflies belonging to 24 genera were recorded in the eighteenth century. Thereafter, no scientific investigations were conducted on the diversity of firefly fauna across Sri Lanka. The present study was conducted to determine the diversity of fireflies of the subfamily Luciolinae in all provinces in Sri Lanka from January to May 2010. Nine *Luciola* species, namely, *L. chinensis*, *L. melaspis*, *L. humeralis*, *L. nicollieri*, *L. candezei*, *L. horni*, *L. intricata*, *L. antennalis* and *L. vespertina* were identified. *Luciola chinensis* was the most widespread and on the whole the most common lucioline firefly in Sri Lanka, but there were regional differences in the dominant species. *Luciola nicollieri* and *L. antennalis* were the most abundant firefly species in Uva province while *L. melapsis* and *L. candezei* were dominant in Sabaragamuwa province. *Luciola chinensis*, however, was the most abundant firefly species in the Southern, Western, Eastern, and Northern provinces, whereas *L. melaspis* was the most abundant firefly species in the Central province, *L. humeralis* in the North Central province and *L. horni* in the North Western province. Climatic variation in different regions of Sri Lanka may influence the dominant species of fireflies in different regions.

Keywords: Firefly, *Luciola*, distribution, Sri Lanka

INTRODUCTION

Taxonomic studies of Sri Lankan fireflies first started in the early 18th century (Wijesekara & Wijesinghe 2003). A survey of a rainforest in 1937 reported 29 firefly species (Baker 1937). In 1966, McDermott reported 30 species of fireflies. Of these, 16 Lucioline species were found in the Central and Northeast Provinces of Sri Lanka. *Luciola melapsis* and *Luciola cingulata* were found in the Southern Province, and an endemic species, *Harmatelia bilineata* (Ototretinae), was found only in the Central Province. The distribution and abundance of fireflies in Sri Lanka is thought to be influenced by climatic variation (Fernando 1984). This paper presents preliminary results of a long term study on fireflies in Sri Lanka.

Firefly species	Sabaragamuwa	Uva	Eastern	North-Western	North-Central	Northern	Central	Western	Southern
L. chinensis (*L. praeusta* complex)	84	116	46	30	52	48	24	32	150
L. melaspis	126	0	12	0	0	0	62	0	3
L. humeralis	0	2	0	0	62	0	0	4	12
L. nicollieri	44	124	0	14	0	0	0	0	0
L. horni	0	0	0	122	0	0	0	0	0
L. antennalis	0	146	38	0	0	24	2	0	0
L. vespertina (*L. praeusta* complex)	48	42	0	0	0	0	0	0	0
L. candezei	112	0	0	0	0	0	0	0	0
L. intricata (*L. praeusta* complex)	0	4	0	0	0	0	0	0	0
Unidentified species 1	4	0	0	0	0	0	0	0	0
Unidentified species 2	2	0	0	0	0	0	0	0	0
Total no. of individuals	420	434	96	166	114	72	88	36	165
Shannon Diversity Index	1.542	1.426	1.033	0.743	0.689	0.636	0.439	0.349	0.289
Evenness Index	0.860	0.686	0.640	0.677	0.594	0.517	0.585	0.504	0.417

Table 1 Number of individuals collected for each *Luciola* firefly species from nine provinces in Sri Lanka. Diversity indices are also shown for each province.

METHODS

The present study was conducted in grasslands in Uva and Sabaragamuwa, and the Central, North-Central, Western, Southern, Eastern, Northern and North-Western provinces from January to May 2010. There were three sampling periods, with each period covering all provinces over five weeks, and with a two week break between sampling periods. Between 6.30 to 9.30 pm, adult fireflies were collected using hand nets, within a 10×10 m^2 quadrat in each provincial site. Most captured fireflies were identified to species, and then released. Individuals that could not be identified in the field were brought back to the laboratory and identified to species by comparing with specimens in the firefly reference collection in the National Museum, Colombo. Data on the numbers of each species of firefly collected in each province were analysed to obtain Shannon's Diversity Index and Evenness Index.

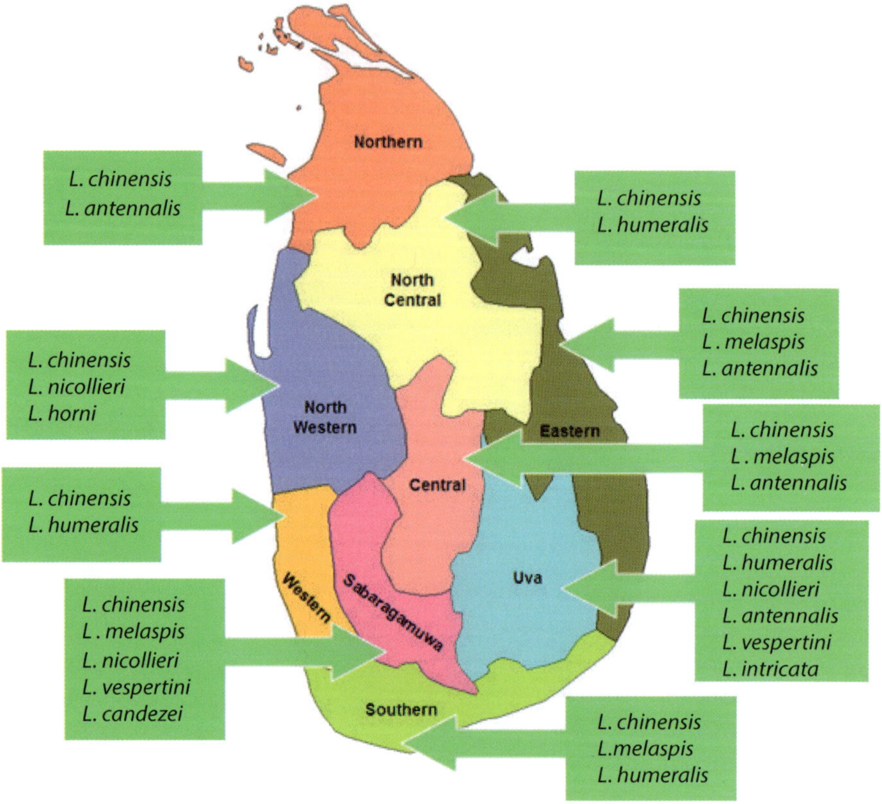

Figure 1 Species of fireflies recorded from nine provinces of Sri Lanka.

RESULTS

A total of 1591 specimens were recorded from grassland habitats in Sri Lanka. Nine *Luciola* species, namely, *L. melaspis*, *L. humeralis*, *L. nicollieri*, *L. candezei*, *L. horni*, *L. antennalis* and 3 species under the *L. praeusta* complex (*L. chinenesis*, *L. vespertina* and *L. intricata*), were tentatively identified during the study.

Luciola chinensis (under the *L. praeusta* complex) was a common species in all nine provinces in Sri Lanka. Uva Province had the greatest abundance of fireflies while Sabaragamuwa Province recorded the highest species richness. *Luciola horni* was collected only from the North-Western province, and *L. candezei* only from the Sabaragamuwa region. *Luciola intricata* was recorded only from Uva Province (Table 1).

Luciola chinensis was the most common Lucioline firefly in Sri Lanka and was the most abundant Lucioline in Southern Province and Uva and Sabaragamuwa provinces. However, dominant species varied according to region.

DISCUSSION

Results of this preliminary study indicate fireflies of the subfamily Luciolinae are common in Sri Lanka. The number of species recorded in this study as occurring in Sri Lanka may be inaccurate as species identifications were carried out in reference to specimens collected during the 18th to early 19th century. The difficulties in identification are complicated by the possibility that some firefly species may have disappeared due to drastic habitat changes occurring after the late 18th century.

Although Sri Lanka is a small island, it has considerable climatic variation due to its geography and monsoon rain patterns. Based on climatic and ecological factors, the country can be divided into three zones: wet, intermediate and dry. The Northern and Eastern Province are situated in the dry zone. The Western Province, Sabaragamuwa and Central Province are situated in the wet zone. The Southern, North-Central and North-Western provinces have parts of all three zones, and Uva Province is situated in the wet and intermediate zone. However, ecological conditions in these provinces are further complicated by topographic conditions, which have considerable effect on rainfall and vegetation patterns. Further study is needed to investigate firefly distribution patterns in relation to the different climatic zones in Sri Lanka.

The abundance of *L. chinensis* (under the *L. praeusta* complex) in the Southern Province, Uva and Sabaragamuwa indicate that *L. chinensis* is well suited to a wet climate. However, *L. chinensis* was rare in the wet Western and Central provinces, possibly because these provinces are highly urbanised. Light pollution and lack of larval food are among factors that may contribute to low populations of *L. chinensis* in these areas. Long term surveys are essential in order to find out the relationship between such environmental factors and the abundance of fireflies.

The abundance of *L. nicollieri* in the colder, wet regions in Uva, Sabaragamuwa and the North-Western Province indicates that this species prefers colder climates with high rainfall. Rainfall may have a major influence on the diversity and abundance of firefly species in Sri Lanka. Long term studies on the distribution patterns and abundance of fireflies in relation to ecological factors is essential in order to understand their ecological requirements and provide a scientific basis for implementing future conservation measures.

REFERENCES

BAKER, J.R. (1937) The Sinharaja rain-forest, Ceylon. *Geographical Journal* **89** 539-551.

FERNANDO, C.H. (Ed.) (1984) *Ecology and Biogeography in Sri Lanka*. 505 pp. Monographiae Biologicae Series, vol. 57. The Hague: W. Junk.

MCDERMOTT, F.A. (1966) Lampyridae. In: *Coleopterorum Catalogus Supplementa, Pars 9* (Ed. by Steel, W.O.), 149 pp. The Hague: W. Junk.

WIJESEKARA, A.P. & WIJESINGHE, D.P. (2003) History of insect collection and review of insect diversity in Sri Lanka. *Ceylon Journal of Science* **31** 43-59.

Lampyrid 2012. 2:142-148

Well managed firefly tourism: a good tool for firefly conservation in Thailand

ANCHANA THANCHAROEN

Department of Entomology, Faculty of Agriculture, Kasetsart University, Ngamwongwan Rd., Jatujak, Bangkok, Thailand 10900.

Email: agrant@ku.ac.th

ABSTRACT In Thailand, fireflies have become a tourism industry, especially in Samut Songkram province. Amphawa district is the most famous destination, where tourists can enjoy the floating market in the late afternoon and a firefly watching trip at night. The synchronous flashing of *Pteroptyx malaccae* on trees along the Mae Klong riverbank is spectacular and fascinating. However, firefly tourism at Amphawa has brought many problems to the local community and also to the firefly population. It provides a case study of how poorly planned tourism can destroy the very thing that attracted tourists. After the problems at Amphawa became obvious, other firefly tourism sites became aware of the cost of tourism and realised the importance of firefly conservation. Knowledge of fireflies and ecotourism management is required to conduct sustainable firefly tourism. In response, Kasetsart University set up service projects to support the efforts of local communities. The activities comprised 1) training people to understand firefly biology and share their knowledge with tourists, 2) providing handbooks to help retain the knowledge and establish knowledge resources within communities and 3) creating signboards at piers for the information of tourists. The results of these services are not only helping to develop firefly ecotourism, but are also creating conservation consciousness in both local communities and tourists. Recently, interest in firefly conservation in Thailand has spread beyond tourism areas. Fireflies are becoming recognised and appreciated as a national treasure of this country.

Keywords: Fireflies, ecotourism, conservation, nature education, Thailand

INTRODUCTION

In the past, there was a high abundance of fireflies in Thailand. The fascinating sight of thousands of synchronous fireflies on mangrove trees along both sides of the Chao Phraya river was well known and was reported more than seventy years ago (Buck & Buck 1976). Besides enjoying the beauty of the flash displays, local people used the flashes for boat navigation at night. Thirty years ago, the firefly populations began to decrease and have disappeared from many areas in Thailand, probably because of urbanisation (Napompeth 2009). The negative effects of urbanisation (habitat destruction and fragmentation, light pollution and water pollution) on firefly populations have also been reported in Malaysia (Nada et al. 2009). With decreasing populations, fireflies are now a rare sight, especially in cities. Firefly flash displays have become an exciting phenomenon to watch, and "firefly tours" have become an important tourist attraction.

Firefly tours in Thailand first started at Plai Pongpang, Samut Songkram province where many traditional Thai-style houses were open for homestay tourists. In 2004, the subdistrict administrative organisation of Amphawa restored the afternoon floating market of Amphawa as a tourist attraction in Samut Songkram and also promoted firefly tours as a nighttime activity. Tourists could take a full day trip, with afternoon activities that continued into the evening. Tourists from all over the country came to Amphawa to watch the impressive synchronous firefly displays. Of the several firefly tour sites in Thailand (Table 1), the Amphawa firefly tours became the most famous, growing quickly from seven tour boats in 2004 to 180 boats in 2010. Without official management, firefly tour operators did not control boat numbers, tourist numbers, or tour times. There were no rules for firefly protection, such as keeping a distance between boats and firefly display trees, limiting light pollution and forbidding the capture or handling of fireflies.

The booming firefly tourism at Amphawa resulted in many changes that affected

Region	Firefly attraction sites	Tour organisers
Central	Mae Klong, Amphawa, Samut Songkhram	Tour operators, resorts
	Chao Phraya, Phrapradaeng, Samutprakarn	Subdistrict administrative organisation of Bangnumpheung
	Bangprakong, Bangprakong, Chasaengsao	Resorts
Eastern	Welu, Klung, Chantaburi	Nature reserve centre
	Klong Proa, Ko Chang, Trad	Restaurant, resorts, local authority
	Ko Krud, Trad	Resorts, tour operators
	Prasae, Klang, Rayong	Resorts
	Bom Rhu canal, Mueng, Chonburi	Local authority
Southern	Khanom canal, Khanom, Nakorn Srithammarat	Restaurant, resorts, tour operators
	Paknakorn, Mueng, Nakorn Srithammarat	Local authority

Table 1 Locations of firefly tour sites in Thailand.

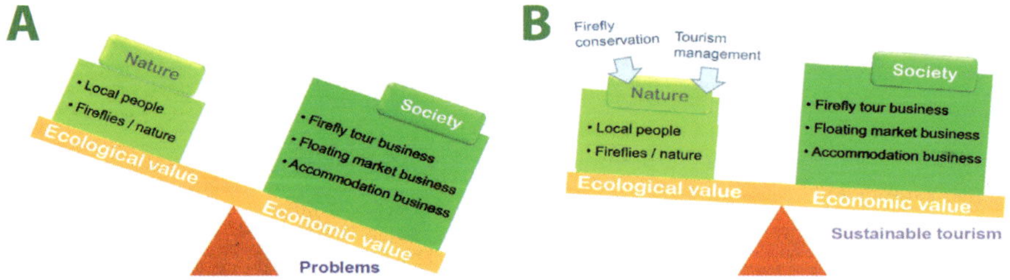

Figure 1 The balance of ecological value and economic value. **A** Problems resulting from over exploitation. **B** Sustainable ecotourism through firefly conservation and tourism management.

the environment and the local community. The many diesel-powered tour boats generated waves that caused riverbank erosion, and also released motor oil that contaminated the riverine system, resulting in a decrease in firefly populations. The construction of new restaurants and resorts destroyed even more firefly habitats. The shift from an agriculture-based community involving mainly riverside residents to a tourism- or business-based community involving boatmen, guides, homestay operators, and souvenir and food sellers, with little benefit to the original residents, led to conflict within the community. Some locals living near the riverbanks cut down firefly display trees to reduce night-time disturbance from the noisy tour boats.

Firefly conservation and tourism management needs to be structured into the firefly tourism activities of Thailand so as to maintain the ecosystem services provided to the people by the river system (Fig. 1). As defined by the UN Millennium Ecosystems Assessment (2005), ecosystem services are human benefits supplied by natural ecosystems in the form of products and processes such as food, water and recreation. The role of nature in continuing human life is well known, and many efforts are being made to maintain biodiversity so as to facilitate a high level of ecosystem services (Jacobsen & Hanley 2009; Mertz et al. 2007; Pearce 2007). To help support sustainable tourism, a project was carried out to impart basic knowledge of fireflies and their conservation to local residents and tour operators. The project was run as an educational service to the communities concerned.

METHODS

The educational service project was conducted from April 2008 to September 2009 at firefly habitats in Khanom and Amphawa in the provinces of Samutprakarn and Nakorn Sri Thammarat, respectively (Table 1; Fig. 2). The project comprised three activities: 1) conducting training programmes, 2) putting up permanent information boards (Fig. 3), and 3) distributing firefly handbooks. The training programmes

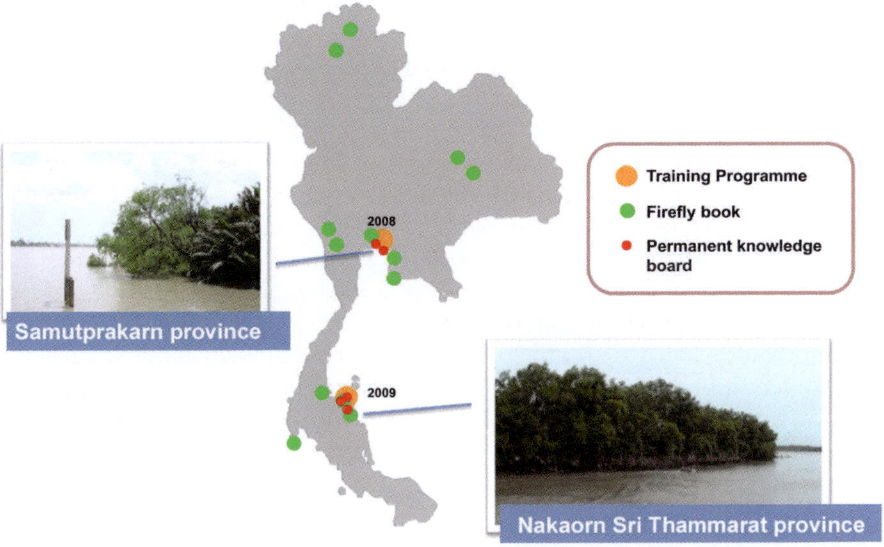

Figure 2 Map of Thailand showing areas where educational service activities were conducted.

were conducted at two existing firefly tourism destinations. A preliminary survey was conducted to find out which aspects of the firefly (e.g., life cycle, ecology, conservation) were of most interest to the locals. A one-day training programme was then planned based on feedback received from the preliminary survey. Participants of the training programme were students, tour guides, boat operators, mangrove protection officers, and local people staying in or near firefly habitats. Each participant was asked to complete a pre-training quiz comprising 15 questions with multiple-choice answers that would gauge basic knowledge on fireflies. Following this, instructional training was given on the biology, ecology, local diversity and conservation of fireflies, as well as on ecotourism in general. A post-training quiz utilising the same questions asked in the pre-training quiz was then conducted to evaluate the effectiveness of the programme. A paired t-test was used to analyse the significance of the difference between pre- and post-training quiz scores.

In addition to the training programme, a total of five permanent signboards were constructed at piers, mangrove visitor centres and recreation areas to impart basic knowledge to tourists and the local community. Information was provided on the species found in the area, their life cycle, aspects of their conservation, and rules to observe when conducting or participating in firefly-watching tours. Also, about 200

Figure 3 Permanent knowledge board at a pier in Samut Prakarn province.

copies of a book containing answers to frequently asked questions about fireflies were distributed in all areas in Thailand that have firefly tours, as a reference that could be used in training programmes and as a resource for public education.

For each project activity, a feedback questionnaire was distributed to participants and visitors to enable them to indicate their satisfaction. A score of 1 to 5 was used with higher scores indicating higher satisfaction. For the training programme, questions related to how well the programme was publicised, duration and level of instruction, quality of instructors, adequacy of the audio-visual teaching aids and overall benefit received. Visitors viewing the signboards were asked to indicate their satisfaction with the choice of location, adequacy of information on fireflies, visual appeal, and overall utility and benefit of the signboards. Questions for recipients of the book included adequacy and reliability of photos and information on fireflies, and overall benefit received.

RESULTS

Twenty six people participated in the preliminary survey. The survey showed that people were most interested in aspects of the general biology of fireflies, particularly the life cycle (26 responses). Respondents were also interested in firefly conservation and eco-tourism (15 and 14 responses, respectively). Firefly taxonomy, breeding and flash behaviour were topics that drew the fewest positive responses (1 to 2 responses).

The training programme was well attended, with 47 and 110 participants at the firefly tourism sites in Phrapadaeng and Khanom, respectively. At Phrapadaeng and Khanom, mean quiz scores prior to training were 10.15 ± 0.41 (n=40) and 8.58 ± 0.24 (n=65), respectively, while mean quiz scores after training were 12.93 ± 0.30 (n=40) and 12.02 ± 0.21 (n=65), respectively. Although approximately 60% of the participants had been on firefly boat tours, their knowledge of the general biology of fireflies was minimal. Many of them understood only the quiz statement that "fireflies are nocturnal insects that can produce light."

Attendees of the training course, visitors viewing the permanent signboard and recipients of the firefly handbook gave scores higher than 4 in the questionnaires, indicating a high level of satisfaction. People gave positive feedback about the three activities conducted and agreed that the activities helped raise awareness about fireflies and boost conservation efforts in the firefly tourism areas.

DISCUSSION

Before firefly tourism increased in Thailand, the decrease in firefly populations attracted little public attention. Firefly tourism has resulted in people becoming aware of and interested in firefly conservation. The wide media coverage of the destruction of firefly habitats in Amphawa has drawn even more attention to the importance of sustainably managing firefly ecotourism. Involved parties have a vested interested in equipping themselves with knowledge on how to run firefly tours in a way that will not harm the fireflies or disturb residents living along the river banks.

For the people in the communities selected for this project, knowledge on fireflies and how to conserve them was gained through the training programme, educational signboards and firefly book. It is hoped that this will pave the way for sustainable firefly tourism with community-led firefly conservation that will benefit both tourists and the local communities while ensuring that the fireflies are protected. Maintaining this balance between ecosystem and economic priorities is important, and since firefly ecotourism would not exist without the fireflies, perhaps firefly conservation efforts could be funded by revenue generated from firefly tours. The tourists can pay for environmental maintenance (Jack et al. 2008), administered through the local authority or government organisations. In Amphawa, where there is no official organisation overseeing firefly ecotourism management, community-led efforts, with firefly tour organisers collaborating to implement these changes, are needed. Also needed country-wide are books, websites and educational videos, so that firefly conservation awareness can keep pace with the rapidly-expanding firefly tourism industry in Thailand.

ACKNOWLEDGEMENTS

I would like to thank the educational service support staff from the Faculty of Agriculture, Kasetsart University. I am also grateful to the local communities and local authorities at the provinces of Samutprakarn (Phrapradaeng) and Nakorn Srithammarat for their generosity and hospitality. Finally, I would like to thank Dr. Steven R. Wing for his comments on an early draft of the manuscript.

REFERENCES

Buck, J. & Buck, E. (1976) Synchronous fireflies. *Scientific American* **5** 74-85.

Jack, B.K., Kousky, C. & Sims, K.R.E. (2008) Designing payments for ecosystem services: Lessons from previous experience with incentive-based mechanisms. *Proceedings of the National Academy of Sciences USA* **105** 9465-9470.

Jacobsen, J.B. & Hanley, N. (2009) Are there income effects on global willingness to pay for biodiversity conservation? *Environmental and Resource Economics* **43** 137-160.

Mertz, O., Ravnborg, H.M., Lövei, G.L., Nielsen, I. & Konijnendijk, C.C. (2007) Ecosystem services and biodiversity in developing countries. *Biodiversity Conservation* **16** 2729-2737.

Millennium Ecosystem Assessment (2005) *Ecosystems and Human Well-being: General Synthesis*. Washington, D.C.: Island Press and World Resources Institute.

Nada, B., Kirton, L.G., Norma-Rashid, Y. & Khoo, V. (2009) Conservation efforts for the synchronous fireflies of the Selangor River in Malaysia. In: *Diversity and Conservation of Fireflies. Proceedings of the International Symposium on Diversity and Conservation of Fireflies, Queen Sirikit Botanic Garden, Chiang Mai, Thailand, 26-30 August 2008* (Ed. by Napompeth, B.), pp. 160-171. Chiang Mai: Queen Sirikit Botanic Garden.

Napompeth, B. (2009) Reminiscience of firefly study in Thailand. In: *Diversity and Conservation of Fireflies. Proceedings of the International Symposium on Diversity and Conservation of Fireflies, Queen Sirikit Botanic Garden, Chiang Mai, Thailand, 26-30 August 2008* (Ed. by Napompeth, B.), pp. 1-10. Chiang Mai: Queen Sirikit Botanic Garden.

Pearce, D. (2007) Do we really care about Biodiversity? *Environmental and Resource Economics* **37** 313-333.

Lampyrid 2012. **2**:149-155

The effect of habitat modification on firefly populations at the Rembau-Linggi estuary, Peninsular Malaysia

Wan Faridah Akmal Wan Jusoh & Nor Rasidah Hashim*

Laboratory of Biodiversity and Environmental Conservation, Faculty of Environmental Studies, Universiti Putra Malaysia, 43400 UPM Serdang, Selangor, Malaysia.

* *Email: rasidah@env.upm.edu.my*

ABSTRACT The synchronously flashing firefly *Pteroptyx tener* congregates on trees along a nine kilometre stretch of the Rembau-Linggi estuary, but their population and the potential economic benefits from firefly ecotourism may be threatened by destruction of mangroves. To investigate the changes taking place to the mangroves on which the fireflies depend, we conducted land-use spatial analysis of satellite images and ground surveys within specific distances from the river. Land use analysis showed that mangroves at the study site have been highly modified over two decades and underwent intensive conversion to other land uses (e.g. agriculture, aquaculture and urbanisation). More recently, canalisation work also destroyed a large area of mangroves. Of the 87 firefly display trees identified two years earlier, only 41 still had fireflies congregating on them, and 14 trees had been destroyed. No new firefly display trees were found. There is an urgent need to conserve the fireflies and their habitat before they become locally extinct.

Keywords: Habitat modification, land-use analysis, GIS, river widening, fireflies, mangroves, Rembau-Linggi

INTRODUCTION

Populations of many riverine firefly species, which occur within the narrow tidal zone along mangrove-lined estuaries in Peninsular Malaysia, have been reported to be on the decline (Nada et al. 2010). In Malaysia, human activities have continued to threaten the habitats of the firefly *Pteroptyx tener* despite the potential economic benefits for the local people from firefly ecotourism, for example in Kuala Selangor (Nada et al. 2010), Kuala Sepetang and Rembau-Linggi (Jusoh et al. 2010a, 2010b). The general perception is that firefly populations in these areas have declined over the years due to habitat destruction and degradation as a result of anthropogenic activities (Ohba 1993; Ohba & Wong 2004). The extent of the decline has not been quantified, but the results of a long term study to monitor the firefly population at Kuala Selangor support these perceptions (Nada et al. 2010).

In the Rembau-Linggi estuary, *P. tener* has been found congregating on several tree species such as *Sonneratia caseolaris* and *Talipariti tiliaceum* growing along a 9 km stretch of the brackish river. The geographic coordinates of the display trees were recorded during a previous survey in 2008 (see Jusoh et al. 2010a). Since then, we have observed a decline in the firefly population that may be related to land use changes taking place in the area. The present study investigated the impact of land use change and habitat modification on the abundance of *Pteroptyx tener* in Rembau-Linggi.

METHODS

Historical and current land use analysis of mangrove buffer zones

Land use maps (1: 50,000 scale) of mangrove areas in the Rembau-Linggi estuary were acquired from the Department of Agriculture, Malaysia for 1984, 1997 and 2006, and scanned at 300 dpi. The scanned images were then digitised in ArcGIS 9.3 to determine land use types in the study area. Zones 100 m, 400 m, 500 m and 1000 m from the riverbanks were delineated in accordance with the Department of Irrigation and Drainage (DID) official buffer zone guidelines on river and river reserve development in Malaysia (DID 2005). Land use types for each zone were identified and the data entered in an attribute table. The area under different categories of land use in different zones was then calculated.

A survey was carried out along the river in June 2010 to record land use changes that had taken place since 2006, referring to the 2006 land use map. Habitat modifications such as newly cleared land and plantation establishment were noted and photographed.

Reassessment of the number of firefly display trees

The locations of the 122 trees that had firefly displays in 2008 were revisited with the aid of a GPS on the night of 12 June 2010 to examine for the presence of fireflies on or near the trees. If firefly displays were no longer present within 50 m of the location of a former display tree, the fireflies were considered to be absent from that location. Any loss or damage of former display trees was also recorded.

RESULTS

Land use attributes for the years 1986, 1997 and 2006 are shown in Table 1. In 1984, mangroves comprised 37.9% of the 1180 ha delineated by the 1000 m buffer zone. Plantations made up 44.6% while other agricultural crops and aquaculture comprised 17.5%. Within the 1000 m buffer zone (i.e., 500-1000 m from the river bank), about 93 ha of mangroves was lost between 1984 and 2006, in part due to

Land use category grouped by distance from riverbank	% Area covered for year 1984	1997	2006	Change (%)
0 - 99 m (260 ha)				
Mangrove forest	76.4	85.8	86.2	9.8
Plantations	20.8	6.5	7.4	-13.4
Other agricultural crops & aquaculture	3.1	3.6	2.8	-0.2
Other forest types	0.0	2.5	1.6	1.6
Urban	0.0	0.0	1.4	1.4
100 - 399 m (730 ha)				
Mangrove forest	62.6	60.0	59.3	-3.3
Plantations	27.8	20.7	21.3	-6.5
Other agricultural crops & aquaculture	8.8	13.3	8.2	-0.6
Other forest types	0.0	4.5	8.9	8.9
Urban	0.0	0.0	0.9	0.9
400 - 499 m (240ha)				
Mangrove forest	48.1	44.2	42.9	-5.2
Plantations	38.9	34.1	32.7	-6.2
Other agricultural crops & aquaculture	10.6	17.3	11.6	0.9
Other forest types	0.0	3.0	11.2	11.2
Urban	0.0	0.0	0.2	0.2
500 - 1000 m (1180 ha)				
Mangrove forest	37.9	31.7	30.0	-7.9
Plantations	44.6	50.9	49.2	4.6
Other agricultural crops & aquaculture	17.5	16.9	8.1	-9.4
Other forest types	0.1	0.4	8.5	8.5
Urban	0.0	0.9	4.9	4.9

Table 1 Land use attributes along the Rembau-Linggi estuary from 1984 to 2006, within 100, 400, 500 and 1000 m of the river banks. 'Plantations' comprises oil palm, rubber and coconut, and includes cleared land; 'other agricultural crops' comprises mixed crops, mixed horticulture, orchards and paddies; 'other forest types' comprises scrub and secondary forest and includes grasslands. 'Change' is the difference between the percentage cover in 2006 and 1984. Percentage subtotals are 100 ± 0.1 due to digitising errors of the system used to scan the maps.

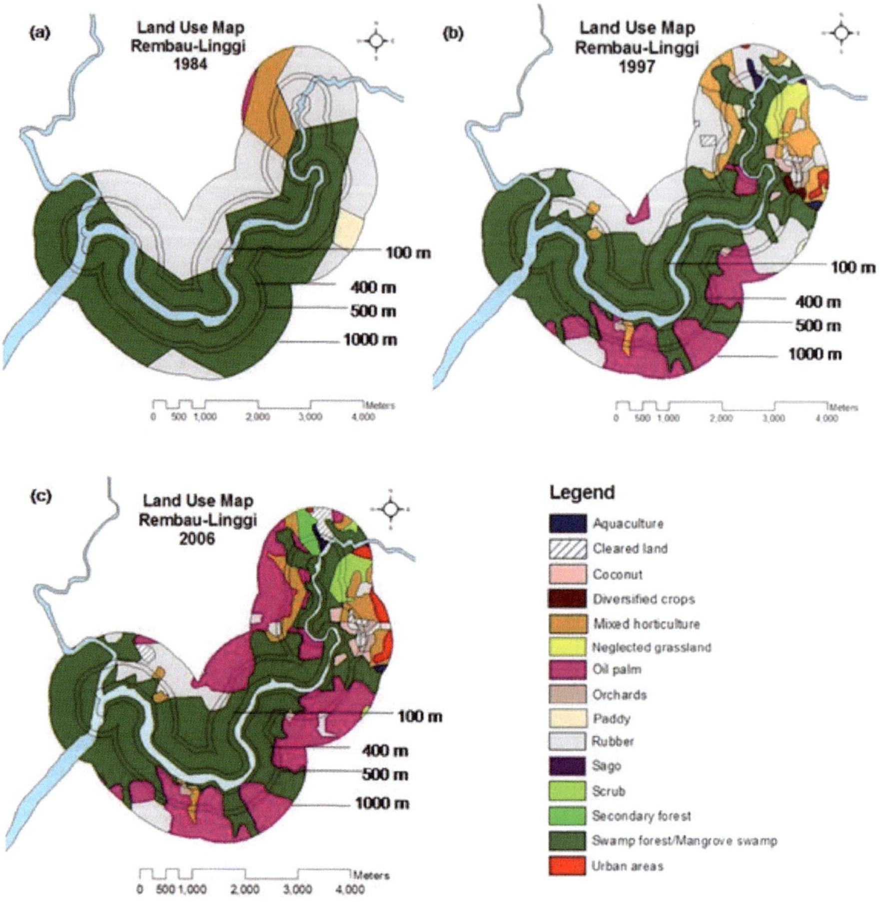

Figure 1 Land-use change maps for an area 1000 m from the riverbanks of the Rembau-Linggi estuary for the years 1984, 1997 and 2006. The lines marked at 100, 400 and 500 m from the riverbanks indicate distance limits for development of tourism, aquaculture, housing and industrial activities, respectively, based on DID guidelines.

land conversion to residential areas, which comprised 4.9% of the land area in 2006. A large area of rubber plantations was converted to oil palm, but the overall area under plantations did not change much, in fact dropping slightly from 50.9% in 1997 to 49.2% by 2006. The area under other agricultural crops (especially paddies) decreased following reversion to scrub and secondary forest, and thus the area under other forest types increased during that time.

Departure from the DID guideline on river and river reserve development occurred in four of the five land use categories. No form of development is permitted

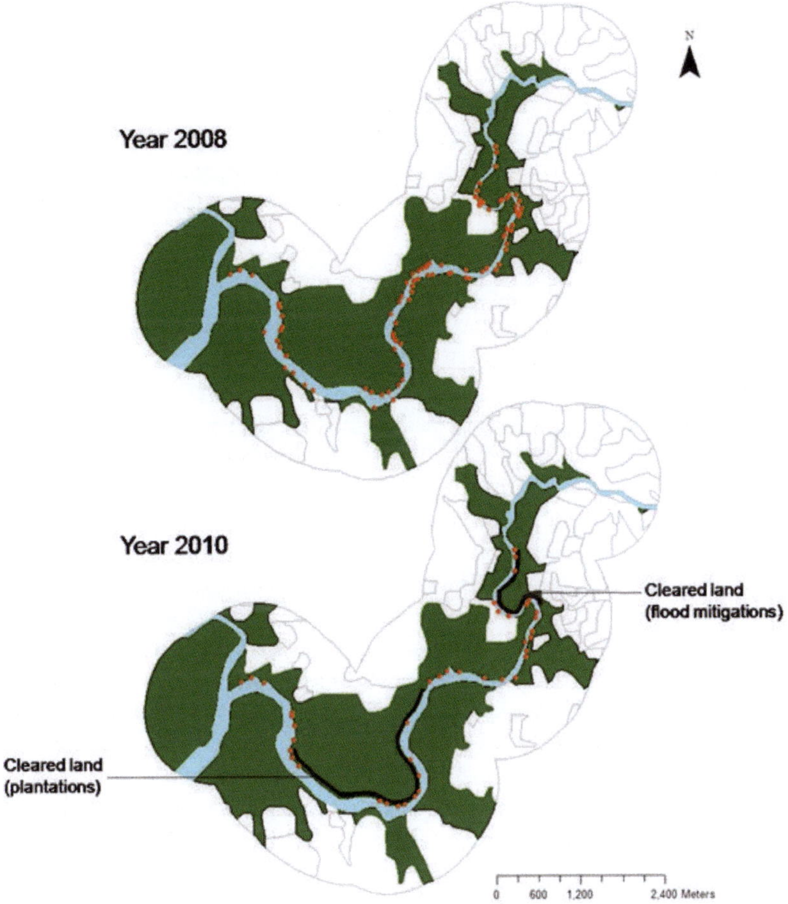

Figure 2 Changes in the number of trees with firefly displays (orange dots) along the riverbanks of the Rembau-Linggi estuary from 2008 to 2010. Black lines show cleared land along the riverbanks. Land use demarcation is based on Figure 1 (2006).

within 100 m of riverbanks, while development for tourism, aquaculture, housing and industrial areas needs to be at least 100, 400, 500 and 1000 m, respectively, from the riverbanks (see DID 2005). In the area assessed, however, aquaculture, which should be more than 400 m away from the river bank, took place within the 400 m buffer zone as well as in the 100 m buffer zone where all forms of development are prohibited (Fig. 1). Similarly, housing development encroached into the 500 m buffer zone of the river bank. Within the 100 m development-prohibited buffer zone, the following were also planted: oil palm, rubber, sago and mixed fruit trees.

Our 2010 ground-truthing survey showed that mangroves within the 100 m buffer zone continue to be cleared to plant oil palm at an increasingly rapid rate. Land-clearing took place right up to the riverbank on long stretches of the river (Fig. 2). Canalisation work (river widening and deepening) carried out by the DID for flood mitigation also cleared swaths of mangroves at least 20 metres wide lining the riverbanks in Linggi and Lubok China. As a result, 14 of the 122 trees that were used by the fireflies for display behaviour when we conducted our study in 2008 had been destroyed by 2010. Within that span of time also, the number of trees populated by fireflies declined by more than half from 122 to 57 (Fig. 2), and no new trees were used for displays.

DISCUSSION

We found that mangrove forest in the Rembau River had been highly modified over a span of two decades and underwent intensive conversion to other land uses (e.g., agriculture, aquaculture and urban areas). This finding is consistent with a report by the Food and Agricultural Organization (FAO 2007), which states that mangrove losses in Malaysia have been primarily due to land use conversion for agriculture, shrimp ponds and urban development. Our land use analysis showed that agriculture, for example plantation crops and subsistence crops, as well as aquaculture, took place within the 100 m buffer zone, while urban development occurred within 400 m.

All of these were encroachments into the buffer zones stipulated under DID guidelines. A major reason for the widespread encroachment was the degazetting of the Rembau-Linggi mangroves as a forest reserve in 2004 (FDPM 2004). This rapidly led to the planting of oil palm and rubber in riparian areas as close as three metres from the riverbank. It is a clear indication that riparian areas in Malaysia receive better protection as part of a forest reserve than they do within buffer zones stipulated by the DID. Riparian buffer zones are neither strictly implemented nor strictly enforced by the state.

Conversion of the mangrove forests to agriculture is bound to endanger the firefly population because its native habitat is destroyed and fragmented. In Thailand, habitat fragmentation is thought to cause the local extinction of fireflies because it limits the range of firefly dispersal (Thancharoen 2007). There may be a similar consequence in the Rembau-Linggi estuary.

It is uncertain if the mangrove forest will recover and how long it will take if it does. If no drastic and immediate steps are taken to protect the fireflies, we fear the population in the Rembau-Linggi mangrove forest will continue to drop and eventually disappear. It is recommended that remaining habitat be protected and the firefly population monitored to help ensure its survival.

ACKNOWLEDGEMENTS

We would like to thank Mohd Fazlin Nazli and Ngatemin Saru for their kind assistance during fieldwork. Grace Tabitha Lim and Laurence Kirton helped improve an earlier version of this manuscript. Nur Azura Adam and Zelina Zaiton Ibrahim shared their expert advice on entomology and mangroves. This study was funded by the Ministry of Higher Education Malaysia, under their Fundamental Research Grant Scheme (FRGS) (vote no. 5524019) and MyBrain15-MyPhD.

REFERENCES

DID (2005) *Garispanduan Pembangunan Melibatkan Sungai dan Rizab Sungai.* [Guidelines for Development Involving Rivers and River Reserves]. Department of Irrigation and Drainage, Kuala Lumpur, Malaysia. (In Malay). http://www.water.gov.my/images/pdf/information_for/garis%20panduan%20pembangunan%20melibatkan%20sungai%20%26%20rezab%20sungai.pdf. Date accessed: 25 September 2010.

FAO (2007) *The World's Mangroves 1980-2005.* Rome: Food and Agriculture Organization (FAO).

FDPM (2004) *Forestry Statistics, Peninsular Malaysia 2004.* Kuala Lumpur, Malaysia: Forestry Department of Peninsular Malaysia.

Jusoh, W.F.A.W., Hashim, N.R. & Ibrahim, Z.Z. (2010a) Distribution and abundance of *Pteroptyx* fireflies in Rembau-Linggi Estuary, Peninsular Malaysia. *Environment Asia* **2** 56-60.

Jusoh, W.F.A.W., Hashim, N.R. & Ibrahim, Z.Z. (2010b) Firefly distribution and abundance on mangrove vegetation assemblages in Sepetang Estuary, Malaysia. *Wetlands Ecology & Management* **18** 367-373.

Nada, B., Khoo, V. & Kirton, L.G. (2010) A flash of hope for the fireflies in Kuala Selangor. In: *FRIM in Focus,* pp. 4-5, December 2010.

Ohba, N. (1993) Preliminary report of the fireflies, *Pteroptyx tener* in Malaysia. *Science Reports of the Yokosuka City Museum* **40** 15.

Ohba, N. & Wong, C.H. (2004) External morphology and ecological study of the firefly, *Pteroptyx tener* at Kampung Kuantan, Selangor, Malaysia. *Science Reports of the Yokosuka City Museum* **51** 1-33.

Thancharoen, A. (2007) *The Biology and Mating Behavior of an Aquatic Firefly Species, Luciola aquatilis sp. nov. Thancharoen (Coleoptera: Lampyridae).* Ph.D. Thesis. Mahidol University.

Lampyrid 2012. **2**:156-161

Mapping riparian vegetation for conservation of a firefly habitat at Kuala Selangor, Malaysia

H. KHALI AZIZ[1]*, O. HAMDAN[1], F. MOHD AZAHARI[1], I. AZHAN SHAH[1], H. MOHD NASIR[1] & N. SAMSU ANNUAR[2]

[1] *Forest Research Institute Malaysia (FRIM), Kepong 52109 Selangor, Malaysia.*
[2] *Selangor Forestry Department, Tingkat 3, Podium Utara, Bangunan Sultan Salahuddin Abdul Aziz Shah, Shah Alam, 40660 Selangor, Malaysia.*

**Email: khali@frim.gov.my*

ABSTRACT Some firefly species are known to have an association with mangrove vegetation. The synchronously-flashing firefly *Pteroptyx tener*, which has become an ecotourism icon of Kuala Selangor in Malaysia, is one such example. These fireflies use *Sonneratia caseolaris* as a favoured display tree, and breed in the riverine vegetation in the flood plains. However, due to over-development, many mangrove areas have been converted to other land uses. Detailed mapping of the extent of mangrove species is important in planning habitat conservation. This paper presents findings from a study to map land use / land cover (LULC) along the Selangor River using remote sensing data, in a 637.1 ha area well-known for its population of *P. tener*. Quickbird satellite images (January 2010) were used to map the distribution of different LULC classes. Nine classes were identified: mixed horticulture, orchards, oil palm plantations, banana plantations, aquaculture, secondary forest, sago palms, nypa palms, and *Sonneratia caseolaris*. *Sonneratia caseolaris* trees comprised 5.3% of land area, but were found all along the riverbank in 10–40 m stretches. Sago palms (*Metroxylon sagu*) and nypa palms (*Nypa fruticans*) comprised 11.2 and 3.0% of land area, respectively. Of the nine LULC classes found in the study area, fireflies are known to be associated with four: *S. caseolaris*, sago palm, nypa palm and secondary forest. Areas with these vegetation types represented only 30.5% of the total land area in the study area and should be considered important conservation sites.

Keywords: Firefly, habitat, mapping, LULC, conservation, GIS, remote sensing, Selangor River, Malaysia

INTRODUCTION

The recent interest in fireflies in Peninsular Malaysia stems from the thriving firefly ecotourism industry, particularly in Kuala Selangor, which is high on the destination list of domestic and foreign tourists (Kirton & Nada 2010; Wong 2009). The firefly species that attracts tourists to Kuala Selangor is *Pteroptyx tener*. It is restricted to mangroves, and each stage of its lifecycle depends on different parts of the mangrove ecosystem (Nada 2011; Nagelkerken et al. 2008; Nada & Kirton 2004; Nallakumar 2004). The adults congregate on mangrove trees at the edges of the Selangor River to display and mate at night. After mating, females fly to muddy banks behind the display trees and lay their eggs in the moist soil. The larvae, which feed on river snails, can be found five to thirty metres from the display trees, in areas where sago palms grow (Nada 2011, Fig. 1). The fireflies are most abundant near Kampung Kuantan, about 15 to 30 km from the mouth of the Selangor River (Murray 1982).

Recently, much of the floodplain vegetation has been cleared to make way for oil palm and banana cultivation. Around 52 ha of the riverine area belonging to a number of private landowners is planned for cash crop cultivation and 32 ha has been cleared as at 2009 (The Star 2009b). Much of the riparian vegetation consisting of nibong (*Oncosperma tigillarium*), sago palm (*Metroxylon sagu*), nypa palm (*Nypa fruticans*) and fig trees (*Ficus* spp.) has been removed, leaving a narrow strip of berembang (*Sonneratia caseolaris*) trees lining the bank. In some parts, the clearing has been conducted right up to the river edges (Nada et al. 2010).

News of the destruction at Kampung Beluntas opposite the visitor centre at Kampung Kuantan was greatly highlighted by the media. In fact, periodic clearing had been occurring since 2007, according to researchers at the Forest Research Insti-

Figure 1 *Sonneratia caseolaris* trees line the riverbanks, with sago palm trees further inland, at a firefly display area on the Selangor River near Kampung Kuantan, Peninsular Malaysia. Photo taken at low tide.

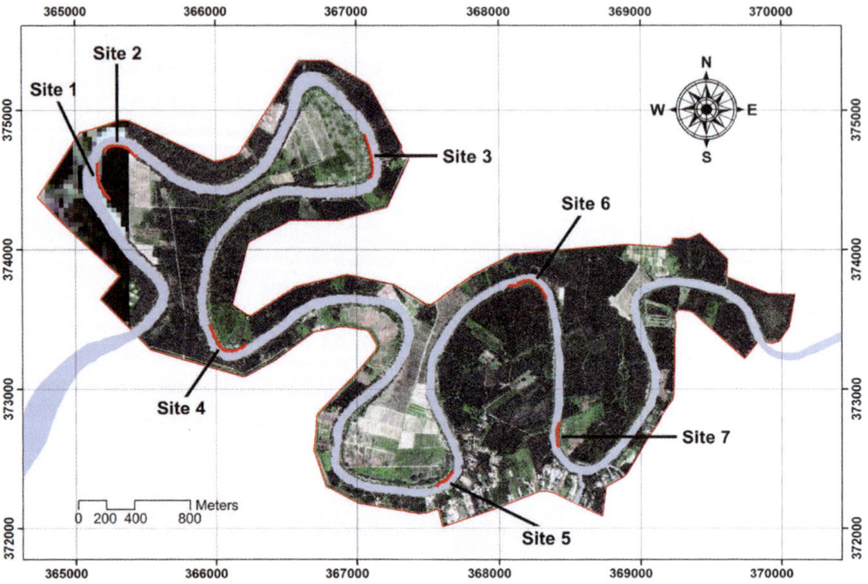

Figure 2 Locations of nine FRIM firefly monitoring stations along a section of the Selangor River near Kampung Kuantan.

tute Malaysia (FRIM) (The Star 2009a). Some parts of this land have been left idle for the last 80 years.

This study was carried out to map firefly habitats in Kampung Kuantan, specifically to identify the types and extents of land use / land cover (LULC) classes as a means identifying areas of conservation importance.

METHODS

Land use / land cover mapping

The study was conducted along the Selangor River near the village of Kampung Kuantan in Kuala Selangor, Malaysia. The study area was confined to about 200 m on each side of a section of the river (from 97.79°N 3.38°E to 97.83°N 3.38°E), which covered 637.1 ha. Land enclosed by the loops of the river was included. Satellite data from QuickBird with a spatial resolution of 0.6 m, acquired on 6 January 2010, was used to classify the LULC of the study area. Image classification was conducted through visual interpretation and manual vectorisation. There were nine LULC classes, namely mixed horticulture, orchards, oil palm plantations, banana plantations, aquaculture, secondary forest, sago palms, nypa palms and *Sonneratia caseo-*

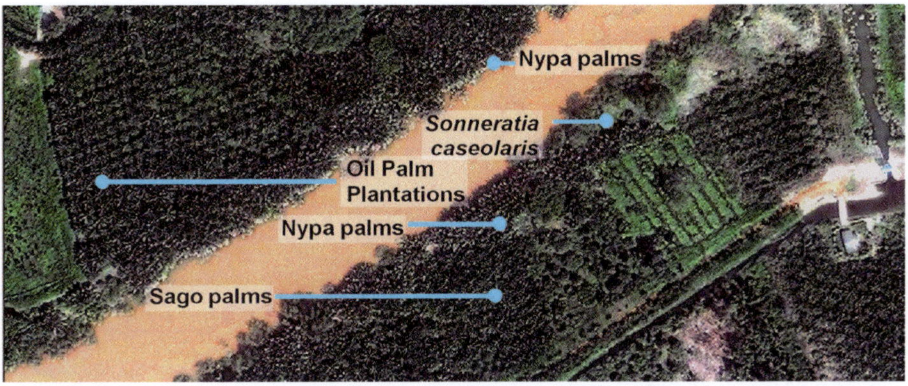

Figure 3 Vegetation types and land cover classes identified on a high resolution Quickbird image.

laris. The image was geo-referenced to the Malaysian Rectified Skew Orthomorphic (MRSO) projection system so that it registered in the same coordinates as ground truthing points.

Field observation

Ground truthing points were collected in the field for all vegetation types to validate the assignment of LULCs on the map. The seven firefly monitoring stations established along the Selangor River in 2006 by the Forest Research Institute Malaysia (FRIM) were overlaid on the map (Fig. 2). The stations were established when a project was initiated to develop a systematic technique for monitoring the firefly population (Kirton et al. 2007, 2012), and they are currently in use for monthly firefly monitoring (Khoo et al. 2012). The stations monitor fireflies on display trees along stretches of riverine vegetation.

Land use / land cover class	%
Mixed horticulture	3.3
Orchards	7.3
Oil palm plantations	32.8
Banana plantations	25.0
Aquaculture	0.9
Secondary forest	11.0
Sago palms	11.2
Nypa palms	3.0
Sonneratia caseolaris	5.3
Total	100

Table 1 Percentage of area under land use / land cover classes in the study area along the Selangor River. Total area analysed was 637.1 ha.

RESULTS

QuickBird high-resolution imaging could be used to successfully delineate small LULC classes and recognise types of riparian vegetation important for fireflies, such as *Sonneratia caseolaris*, nypa palms and sago palms (Fig. 3).

Sonneratia caseolaris trees (berembang) comprised only 5.3% of land area (Table 1), but this type of vegetation was found all along the riverbank, in 10–40 m stretches (Fig. 4). The

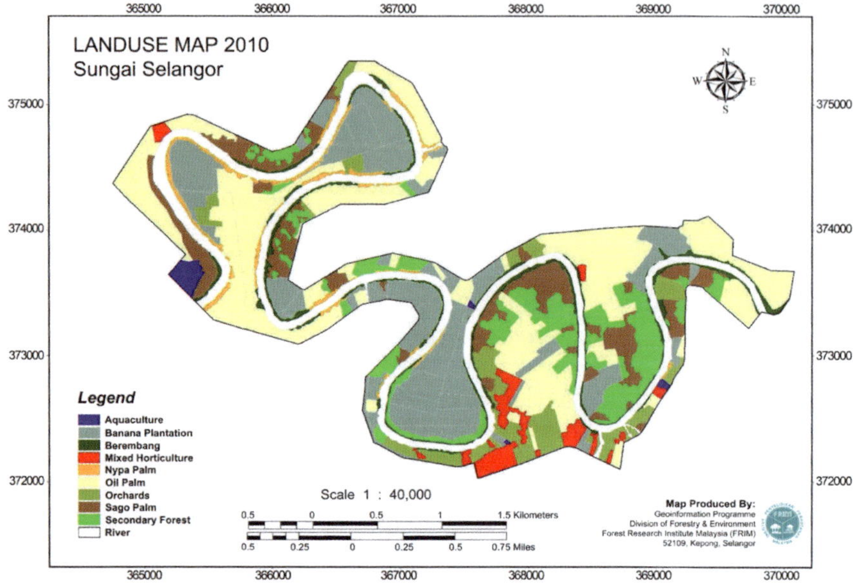

Figure 4 Land use / land cover types in the year 2010, along a section of the Selangor River inhabited by fireflies.

sago and nypa palms found along the riverbank comprised 11.2 and 3.0% of land area, respectively.

Active land clearing along the river, probably for banana and oil palm plantations, was seen when conducting field observations.

DISCUSSION

The study shows that remote sensing data can be useful in mapping vegetation types important as firefly habitats. Information on LULC classes, their specific locations and their extent can be evaluated in relation to existing firefly display areas to better understand the association between firefly habitats and vegetation types. Of the nine LULC classes found in the study area, fireflies are known to be associated with *Sonneratia caseolaris*, sago palm, nypa palm and secondary forest, which comprised only 30.5% of land cover in the study area. These riparian vegetation types are used by the fireflies either as display trees or breeding grounds (Nada 2011; Nagelkerken et al. 2008; Nada & Kirton 2004), and thus should be conserved to protect the fireflies. The state government is making an effort to gazette areas identified as firefly habitats as protected zones (Nada et al. 2009). This move may ensure the long term survival of firefly populations in the area.

REFERENCES

Khoo, V., Nada, B., Kirton, L.G. & Phon, C.-K. (2012) Monitoring the population of the firefly *Pteroptyx tener* along the Selangor River, Malaysia for conservation and sustainable ecotourism. In: *Firefly Conservation: From Science to Practice. Proceedings of the Second International Firefly Symposium, Selangor, Malaysia, 2-5 August 2010* (Ed. by Kirton, L.G., Lim, G.T. & Day, J.C.) *Lampyrid* **2** 162–173.

Kirton, L.G., Nada, B., Khoo, V. & Phon, C.-K. (2012) Monitoring populations of bioluminescent organisms using digital night photography and image analysis: a case study of the fireflies of the Selangor River, Malaysia. *Insect Conservation and Diversity* **5** 244–250.

Kirton, L.G., Nada, B. & Cheng, S. (2007) Keeping the river glowing: Development of a monitoring programme for the fireflies of Sungai Selangor. In: *Sungai Selangor: State-of-the-River 2006* (Ed. by Department of Irrigation and Drainage Malaysia), pp. 46-52. Kuala Lumpur: Department of Irrigation and Drainage.

Kirton, L.G. & Nada, B. (2010) Firefly ecotourism in Malaysia. In: *World Atlas of Mangroves* (Ed. by Spalding, M., Kainuma, M. & Collins, L.), p. 115. London: Earthscan Ltd.

Murray, M.G. (1982) The fireflies of Kuala Selangor. *Malayan Naturalist* **35** 37–38.

Nada, B. (2011) *Ecology and Behaviour of Fireflies (Coleoptera: Lampyridae) Along the Selangor River, Kuala Selangor, Peninsular Malaysia.* M.Sc. Thesis. University of Malaya.

Nada, B. & Kirton, L.G. (2004) The secret life of fireflies. *IRBM Updates*, No. 3, pp. 2–4.

Nada, B., Khoo, V. & Kirton, L.G. (2010) A flash of hope for the fireflies of Kuala Selangor. *FRIM in Focus*, pp. 4-5, December 2010.

Nada, B., Kirton, L.G., Norma-Rashid, Y. & Khoo, V. (2009) Conservation efforts for the synchronous fireflies of the Selangor River in Malaysia. In: *Diversity and Conservation of Fireflies. Proceedings of the International Symposium on Diversity and Conservation of Fireflies, Queen Sirikit Botanic Garden, Chiang Mai, Thailand, 26-30 August 2008* (Ed. by Napompeth, B.), pp. 160-171. Chiang Mai: Queen Sirikit Botanic Garden.

Nagelkerken, I., Blaber, S.J.M., Bouillon, S., Green, P., Haywood, M., Kirton, L.G., Meynecke, J.-O., Pawlik, J., Penrose, H.M., Sasekumar, A. & Somerfield, P.J. (2008) The habitat function of mangroves for terrestrial and marine fauna: A review. *Aquatic Botany* **89** 155–185.

Nallakumar, K. (2004) *Management and Conservation of Fireflies in Peninsular Malaysia.* Ph.D. Thesis. University of London.

The Star. (2009a) Tuesday March 10, 2009. Fireflies lose sparkle as mangroves come down. Star Publications (M) Bhd. http://thestar.com.my/metro/story.asp?file=/2009/3/10/central/3416102&sec=central. Accessed 24th June 2011.

The Star. (2009b) Tuesday April 21, 2009. Habitat loss in Kuala Selangor's firefly colony. Star Publications (M) Bhd. http://www.thestar.com.my/lifestyle/story.asp?sec=lifefocus&file=/2009/4/21/lifefocus/3616808. Accessed 24th June 2011.

Wong, C.H. (2009) Firefly watching and conservation involving local communities in Malaysia. In: *Diversity and Conservation of Fireflies. Proceedings of the International Symposium on Diversity and Conservation of Fireflies, Queen Sirikit Botanic Garden, Chiang Mai, Thailand, 26-30 August 2008* (Ed. by Napompeth, B.), pp. 141-148. Chiang Mai: Queen Sirikit Botanic Garden.

Lampyrid 2012. **2**:162-173

Monitoring the population of the firefly *Pteroptyx tener* along the Selangor River, Malaysia for conservation and sustainable ecotourism

Veronica Khoo*, B. Nada, Laurence G. Kirton & Chooi-Khim Phon

Tropical Forest Biodiversity Centre, Forest Research Institute Malaysia, 52109 Kepong, Selangor, Malaysia.
**Email: veronica@frim.gov.my*

ABSTRACT The Selangor River estuary is well known for its large population of synchronously flashing fireflies, *Pteroptyx tener*, which congregate mainly on *Sonneratia caseolaris* growing on the edges of the riverbank. As a result, ecotourism in the coastal town of Kuala Selangor and its neighbouring villages has become an important contributer to the livelihood of the local communities. In recent years, there has been thought to be a decrease in the numbers of fireflies due to factors such as river pollution, water extraction upstream and clearing of vegetation. As part of conservation efforts for the firefly, a monitoring programme was initiated in 2006. Using digital night photography and image analysis to obtain a count of firefly flashes, the adult firefly population was monitored in seven sites totalling 1.6 km of the riverbank. Three years of data from May 2006 to April 2009 are presented and compared against rainfall, relative humidity, temperature, air quality, water quality and water salinity measurements. Peak adult firefly abundance occurred between June and August, with a secondary but less distinct peak occurring between December and February. The primary peak was distinct in the first two years, but lower and more spread out in the third year, possibly due to changes in rainfall patterns or loss of breeding habitat behind the display trees. Detection of habitat loss led to action by government authorities to help conserve riverside vegetation, which has been an important outcome of the project. The importance of long-term monitoring of the population is discussed.

Keywords: Firefly, *Pteroptyx tener*, population, monitoring, conservation, Selangor River, Malaysia

INTRODUCTION

The small town of Kuala Selangor in Malaysia is well known for firefly ecotourism (Wong 2009; Kirton & Nada 2010). During the night, large numbers of the synchronously flashing firefly *Pteroptyx tener* congregate on *Sonneratia caseolaris* trees

(Mangrove Apple, locally called Berembang) along the banks of the Selangor River near Kuala Selangor (Nallakumar 2002). Inland of the riverbanks, natural and naturalised stands of sago palms (*Metroxylon sagu*), nibong (*Oncosperma tigillarium*), figs (*Ficus* spp.) and betel nut palms (*Areca catechu*) form the habitat of the firefly larvae and their host snail *Cyclotropis carinata* (Nada et al. 2009a). Ecotourism in Kuala Selangor and its neighbouring villages such as Kampung Kuantan has become an important source of income for local communities.

In recent years, however, villagers as well as visitors have felt a decline in firefly numbers. The perceived decline was highlighted by the local media (Nada et al. 2007). There are several factors that could threaten the firefly population, such as river pollution, water extraction upstream, clearing of vegetation and erosion. With rapid development taking place in the area, loss of riverside vegetation as a result of land conversion to agriculture is a major threat to the firefly population (Nada & Kirton 2004; Nada et al. 2009b).

A successful monitoring programme is essential for detecting change, especially when there are threats to a population or when action is taken to restore a population (Hellawell 1991). Realising the need for scientific data on the firefly population, the Forest Research Institute Malaysia (FRIM) was commissioned by the governments of Malaysia and Denmark to develop a monitoring technique and monitoring programme for the firefly population under the Integrated River Basin Management (IRBM) Project (Kirton et al. 2007). The technique developed involves digital photography and image analysis. The results of the first year's monitoring (May 2006 to May 2007) were presented at the National Integrated River Basin Management Seminar in Langkawi Island in 2007 (Nada et al. 2007, 2012). The present paper reports and analyses three years of monitoring data, combining the first year's data with two years of additional data from June 2007 to April 2009.

METHODS

Monitoring was carried out using digital night photography and image analysis. Details of the technique have been described in a separate paper (Kirton et al. 2012). Images captured were analysed using image analysis software to obtain counts of bright spots (firefly flashes) that were calibrated against visual (manual) counts. The counts were used as an index of abundance of the firefly population. This non-destructive method enabled large stretches of Berembang trees to be monitored, thus decreasing the likelihood of the results being influenced by changes in favoured display trees.

Figure 1 shows the locations of the seven monitoring sites (c.f. Nada et al. 2007, 2012). The seven sites were selected based on criteria such as the availability of a wide and continuous stretch of berembang trees where firefly aggregations could be seen and a clear and complete view of the area being monitored. Sites were spread out along the Selangor River, between the villages of Kampung Sepakat and Lubuk

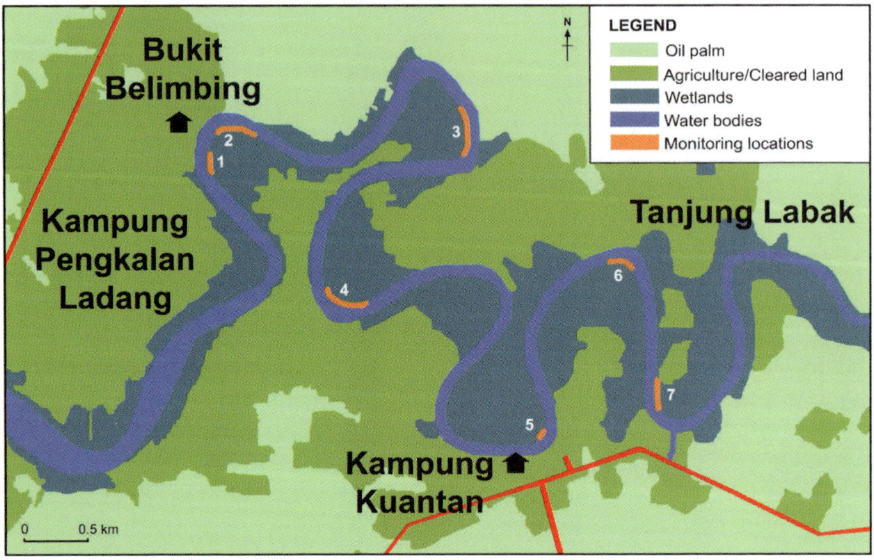

Figure 1 Locations of the seven sites monitored for firefly abundance along the Selangor River (numbered 1 to 7). Original map source: Federal Department of Town and Country Planning (DTCP), Peninsular Malaysia (Land-use in 2006).

Jaya, as this stretch of the river has been said to be the primary area in which the fireflies aggregate (MNS 2002). Table 1 gives further details of sites. The combined length of the stretches of berembang trees being monitored was approximately 1.6 km.

The monitoring exercise was carried out monthly over a three night period, centring on the new moon, with an allowance of one day earlier or later. The same site order was used each month. Each site had three sets (replicates) of images captured for the whole stretch of display trees monitored. The time lapse between each replicate was approximately one minute. Three replicates have been shown to adequately reduce the effects of variation in flash rhythms (Kirton et al. 2012).

The camera was placed on a tripod on one side of the riverbank and used to photograph, section by section, a fixed stretch of display trees on the opposite side of the river by panning the camera through a preset series of angles. The left sides of the images were then cropped for overlap. Camera location and orientation were maintained by having a permanent ground marker and plumb line for the tripod, as well as by using the tripod's graduated dials or a camera-mounted compass.

Most images were captured using a Canon EOS 5D with an EF 70-200 mm f/2.8L IS USM zoom lens. The camera was pre-focused during the day, and at night the camera was set to an equivalent of ISO 3200, with f. 2.8 aperture and 0.5 s exposure

Site*	Range of distances from camera to display trees (m)	Lens focal length setting (mm)	Total length of display trees monitored (m)†
1	93 – 116	135	121
2	74 – 168	135	265
3A	76 – 272	135	369
3B	71 – 100	135	
4A	69 – 181	115	225
4B	74 – 140	135	
5	59 – 88	135	112
6	58 – 190	85	309
7	62 – 123	125	220
TOTAL			**1621**

*Site numbers with the suffix 'A' or 'B' had two camera locations per site to enable a longer stretch of display trees to be photographed.
† The values may not add up to the total due to rounding.

Table 1 Camera-subject distances, focal length and total length of display trees monitored in each site.

time. Depending on sites, the focal length ranged from 85 to 135 mm (Table 1). For the months of May and June 2006 (at the start of the monitoring programme), images were captured using a Nikon D70 with an AF-S DX Zoom Nikkor 18-70 mm f/3.5-4.5G IF-ED zoom lens, at maximum focal length (70 mm), minimum aperture (f. 4.5) and a slow speed of 30 s. There may be a higher margin of error for the counts from May to July 2006 and for January 2007 as they were adjusted to account for the use of a different camera system in May and June, a different trial exposure time in July and incorrect focus in January (Nada et al. 2007, 2012). Adjustments were made based on correlations with counts taken with the camera system and settings in use for the other months.

The index of abundance was compared with environmental measurements of weather, river water quality and air quality, where possible up to four months before the commencement of population monitoring. Comparisons between the monthly firefly population index and environmental parameters were made by graphing the three years of data and visually inspecting the trends. Data for environmental parameters were obtained from the Meteorological Department and Department of Environment (DOE). Average daily rainfall for each month was obtained by averaging data from three weather stations, namely Sungai Buloh Estate Station, Selangor River Station and MARDI Tanjong Karang Station, which were 7.9 km, 4.2 km and 21.7 km from Kampung Kuantan, respectively. Temperature and relative

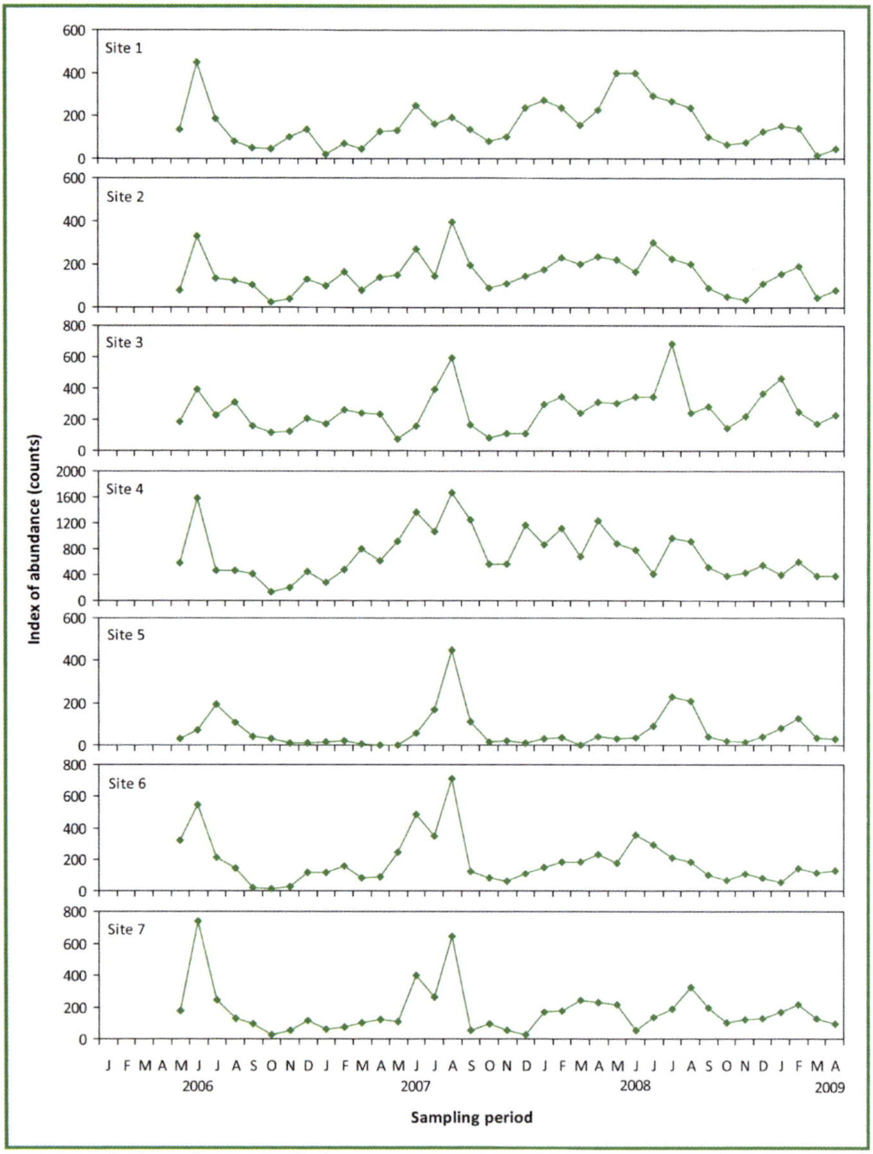

Figure 2 Monthly indices of firefly abundance from May 2006 to April 2009 at seven monitoring sites along the Selangor River. Open symbols represent months for which data was adjusted by calibration (see text for explanation). Due to the timing of the new moon, there were two monitoring periods in the month of July 2008.

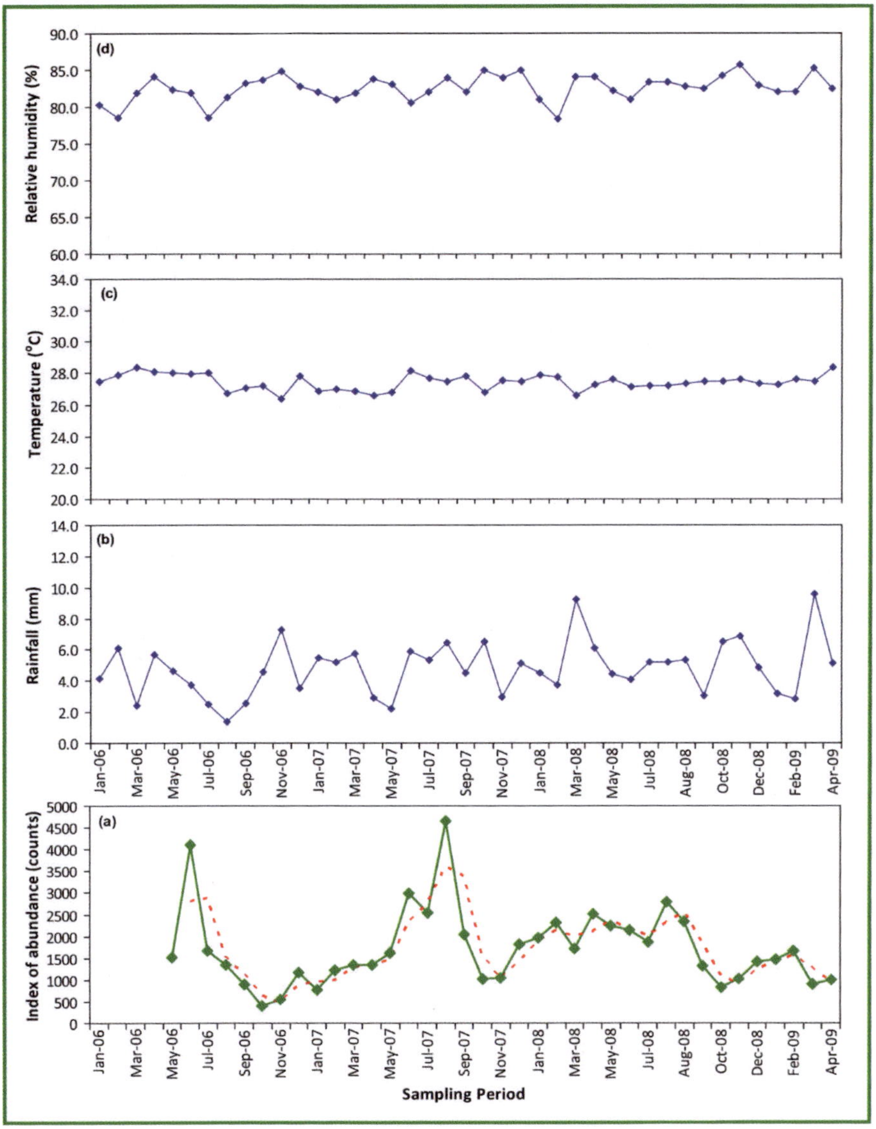

Figure 3 Comparison of monthly firefly abundance at the Selangor River with monthly climatic measurements. (a) Combined monthly firefly population index (total for all sites) from May 2006 to April 2009 (solid line) and a two-month-interval moving average (dotted line), (b) average daily rainfall for each month, (c) average monthly temperature, and (d) average monthly relative humidity. There were two monitoring periods in July 2008 due to the occurrence of the new moon twice in that month, and therefore the climatic measurements for July are repeated in their respective graphs.

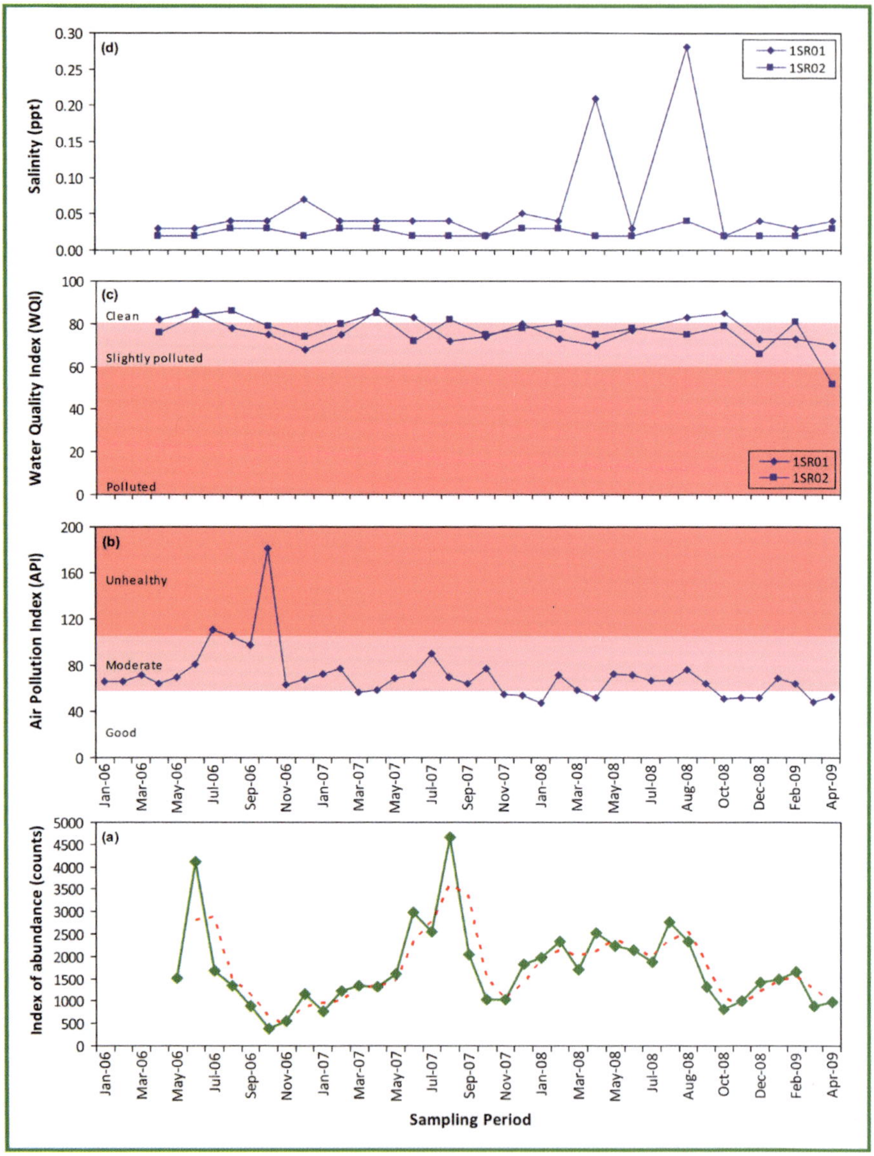

Figure 4 Comparison of monthly firefly abundance at the Selangor River with monthly river water and air quality measurements. (a) Combined monthly firefly population index (total for all sites) from May 2006 to April 2009 (solid line) and a two-month-interval moving average (dotted line), (b) maximum air pollution index (API) reached in month, (c) water quality index (WQI), and (d) water salinity. See legend of Figure 3 for clarification of July 2008.

humidity (RH) data were available only from the MARDI Tanjong Karang Station. Average monthly temperature and relative humidity were obtained for the daily averages of the 8.00 am and 2.00 pm readings. For the salinity and water quality index (WQI), the DOE provided bimonthly readings from two stations, namely 1SR01 near Kampung Kuantan and 1SR02 at a bridge near Batang Berjuntai. Air quality data was obtained from a station located near Bukit Belimbing 5.2 km from Kampung Kuantan, which took two daily readings at 11.00 am and 5.00 pm. The monthly maximum value for the air pollution index (API) was used for the purpose of this study.

RESULTS

In general, firefly abundance at all sites peaked around the middle of the year. In fact, all sites except one peaked between June to August every year, and the latter peaked only a month earlier in May (Fig. 2). In 2006, all of the sites had the highest firefly abundance in June with the exception of Site 5, which peaked a month later. In 2007, the highest abundance was in August except in Site 1, which had its peak in June. In 2008, the timing of peak abundance varied between sites, occurring from April to August. In addition to the mid-year peaks, secondary but less distinct peaks occurred between December and February each year.

The overall monthly index of abundance for all seven sites combined is shown in relation to rainfall, temperature and relative humidity in Figure 3, and in relation to air quality, water quality and salinity in Figure 4. The dotted line on the graph shows a moving average based on an interval of two months. From this graph, the overall peak abundance for each year could be seen to be around the middle of the year, between June and August. The peak for the year 2008 was about 30–40% lower than that of the previous two years. However, the 2008 peak was also noticeably more spread out.

The average monthly temperature and relative humidity (Fig. 3) was relatively stable and varied little throughout the three year period. It ranged from 26.4–28.4°C for temperature and from 78.3–85.7% for RH. Rainfall was not distinctly seasonal but varied between months (Fig. 3). There were periods of lower rainfall in August 2006, May 2007 and February 2009, and periods of higher rainfall around February 2006, November 2006, August 2007, March 2008, November 2008 and March 2009.

Water salinity, as measured at the DOE stations along the river, was generally low, usually remaining under 0.05 ppt (Fig. 4). However, on two months, salinity readings rose to between 0.2–0.3 ppt. Water quality was relatively stable, with the WQI ranging from 52–86 (Fig. 4) and generally in the clean and slightly polluted categories. Air quality readings were usually in the moderate range, except from July to October 2006, which recorded APIs close to or well into the unhealthy range due to transborder haze from forest fires (Fig. 4). October had the highest reading at 181.

The images taken through the monthly firefly monitoring programme also provided evidence of habitat loss, which could be seen in changes in the tree line in images. Habitat loss as a result of riverside vegetation clearing to make way for agriculture was first detected in images taken in mid-2007. Between then and April 2009, a further site was cleared each year. Thus, a total of four out of the seven sites monitored were affected by land clearing for agriculture during the three-year monitoring period.

DISCUSSION

Temperature and relative humidity, which were relatively stable, did not appear to influence trends in the abundance of adult fireflies. In addition, slight drops in water quality in December 2006 and December 2008 appeared to have no affect on the population of adult fireflies in subsequent months.

The increases in river water salinity in two of the months may have been due to differences in time of measurement in relation to tides and water flow rates. Water flow is regulated by an upstream dam and barrage. Salinity in the Selangor River fluctuates with tides and water flow, being lower during low tides and periods of high water flow, and higher at high tides and periods of low water flow (Hashim & Hassan 2012). Salinity at Kampung Kuantan is known to sometimes be within an undetectable range or be up to 0.1 ppt or higher (MNS 2002), with levels of 2.1 ppt having also been recorded (Hashim & Hassan 2012). The spikes in salinity in the DOE readings were, therefore, not beyond what is normally encountered in Kampung Kuantan, and are unlikely to have had any affect on the firefly population. Berembang is said to tolerate long-term salinities of up to 0.5 ppt (MNS 2002).

The noticeably low firefly population in October 2006 has been attributed by villagers and the media to poor air quality due to transborder haze, which began in July 2006 and peaked in October 2006 (Nada et al. 2007, 2012). However, for three consecutive years from 2006-2008, the month of October had the lowest population levels, regardless of air quality levels. Air quality alone, therefore, does not explain the low population in this month. It may however, have been a contributing factor in that year, as the population was at its lowest in October 2006.

For the first year's monitoring (May 2006 to May 2007), it was suggested that high rainfall from November to December and March to May could have caused the population increase around the middle of the year, while low rainfall in the middle of the year could have caused the decrease in population from September to November (Nada et al. 2007, 2012). The basis was that wet conditions may favour larval development and the host snail population. It is thought that the life cycle of *Pteroptyx tener* takes approximately 3 months (Nada & Kirton 2004), which would roughly correspond with the lapse between rainfall conditions and consequent adult firefly abundance. While periods of high and low rainfall do appear to be associated with subsequent adult firefly populations from mid 2006 to late 2007, the pattern is less clear from then on till April 2009. The population did not reach a clear peak in this period, and secondary peaks in the early part of the year seemed to be higher than before, merging with a mid-year peak that was lower and more spread out than usual. This may be due to less marked periods of low and high rainfall, but could also be due to changes in the habitat.

Loss of habitat behind monitored sites was detected as early as 2007, with two sites having been cleared of all but the river-edge display trees (Nada et al. 2007, 2012). By April 2009, a further two sites were cleared leaving only three of the seven sites with riverside vegetation behind the display trees. Loss of natural habitat has been said to be the single most important factor threatening the firefly population along the Selangor River (Nada et al. 2007, 2012), and has been taking place at a rapid rate (Khoo et al. 2009). An analysis of land use / land cover patterns based on satellite images taken in January 2010 along the stretch of river inhabited by fireflies shows that about a quarter of land area remains under natural and naturalised vegetation cover while more than two thirds has been converted for agriculture, primarily oil palm (Khali Aziz et al. 2012). It is possible the change in firefly population trends in 2008 onwards, with a lower and more spread out peak, is due to a decrease in available habitat for the firefly larvae and their host snail. The larvae and host snail inhabit riverside vegetation behind the display trees, and are scarcely found in areas converted for agriculture (Nada et al. 2009a).

Loss of habitat was reported by FRIM to the Kuala Selangor District Council and the Selangor state government in January 2009 and subsequently brought to the attention of the local media, which greatly highlighted the issue. As a result, sections of the Selangor River and its banks in the District of Pasangan, Kuala Selangor have been gazetted, effective 2 July 2009, as a protection zone under Section 48 of the Selangor Waters Management Authority Enactment 1999. Under this enactment, private land owners are not allowed to clear natural vegetation from protection zones. Efforts are also being made to reacquire affected land for rehabilitation.

CONCLUSION

Population trends for the third year of monitoring differed from the first two years, with a lower and more spread-out peak. Possible reasons for this are changes in rainfall patterns and loss of breeding habitat. Habitat loss was detected each year during the monitoring period. However, longer-term monitoring will be required to determine whether or not the changes in population trends are transient and to quantify the actual effects of habitat loss on the population. The detection of riverside vegetation-clearing and the subsequent action taken by government authorities to help protect the habitat of the firefly have been important outcomes of the monitoring project that demonstrate the importance of an ongoing monitoring programme to aid in the conservation of iconic flagship species.

ACKNOWLEDGEMENTS

Funding for developing and conducting a monitoring programme for the fireflies of the Selangor River was provided by the Malaysian Department of Irrigation and Drainage (DID) in cooperation with the Danish International Development Agency (DANIDA) from 2006 to 2007, and by the Malaysian Economic Planning Unit in the Prime Minister's Department from 2008 to 2010. We are grateful to the Selangor State Government and Selangor Waters Management Authority for their interest and cooperation in addressing habitat loss issues. Mr. A.R. Muhammad Farid (FRIM) provided valuable assistance in mapping and measuring sites. We extend our thanks to colleagues in the Entomology Branch of FRIM, especially Mr. M.S. Shaiful Amri and Mr. O. Khirul Faizal, who greatly assisted in field work and data collection. Last but not least, we would like to thank the villagers of Kampung Kuantan, especially Mr. J. Abdul Razak, for their cooperation and help.

REFERENCES

Hashim, N. & Hassan, A.J. (2012) Salinity intrusion modeling for Sungai Selangor. NARBO Newsletter contribution. Network of Asian River Basin Organisations (NARBO). http://www.narbo.jp/narbo/contribution/contrib_003.htm. Accessed: 9 July 2012.

Hellawell, J.M. (1991) Development of a rationale for monitoring. In: *Monitoring for Conservation and Ecology* (Ed. by Goldsmith, B.), pp. 1–14. London: Chapman and Hall.

Khali Aziz, H., Hamdan, O., Mohd Azahari, F., Azhan Shah, I., Mohd Nasir, H. & Samsu Annuar, N. (2012) Mapping riparian vegetation for conservation of a firefly habitat at Kuala Selangor, Malaysia. In: *Firefly Conservation: From Science to Practice. Proceedings of the Second International Firefly Symposium, Selangor, Malaysia, 2–5 August 2010* (Ed. by Kirton, L.G., Lim, G.T. & Day, J.C.). *Lampyrid* **2** 156–161.

Khoo, V., Kirton, L.G. & Nada, B. (2009) The fate of *Pteroptyx tener*. In: *Conservation Malaysia*, pp. 5–6, Issue No. 10, September 2009.

Kirton, L.G. & Nada, B. (2010) Firefly ecotourism in Malaysia. In: *World Atlas of Mangroves* (Ed. by Spalding, M., Kainuma, M. & Collins, L.), pp. 115. London: Earthscan.

Kirton, L.G., Nada, B. & Cheng, S. (2007) Keeping the river glowing: Development of a monitoring programme for the fireflies of Sungai Selangor. In: *Sungai Selangor State-of-the-River Report 2006*, pp. 46–52. Kuala Lumpur: Department of Irrigation and Drainage (Malaysia).

Kirton, L.G., Nada, B., Khoo, V. & Phon, C.-K. (2012) Monitoring populations of bioluminescent organisms using digital night photography and image analysis: A case study of the fireflies of the Selangor River, Malaysia. *Insect Conservation and Diversity* **5** 244–250 (with 4 pages of online supporting information). First published online 20 July 2011, doi: 10.1111/j.1752-4598.2011.00157.x.

MNS (Malaysian Nature Society). (2002) *Firefly Studies in Kuala Selangor District—Final Report Submitted to Selangor State Government.* Kuala Lumpur: Malaysian Nature Society. (Comprises 7 working papers, appendices and an introductory section of 25 pp).

Nada, B. & Kirton, L.G. (2004) The secret life of fireflies. In: *IRBM Updates. Integrated River Basin Management,* pp. 2–4, Issue No. 3, December 2004.

Nada, B., Kirton, L.G., Cheng, S., Shahlinney, L. & Phon, C.-K. (2007) A special case: Monitoring the fireflies of the Selangor River. In: *Reports on Integrated River Basin Management: The IRBM Project—A Joint Effort by Malaysia and Denmark.* Paper presented at the National Integrated River Basin Management Seminar, Langkawi Island, 3–5 July 2007. Published electronically on CD-ROM.

Nada, B., Kirton, L.G. & Norma-Rashid, Y. (2009a) The fireflies of Kuala Selangor: Conservation of berembang alone won't save them. In: *Conservation Malaysia*, pp. 4–5, Issue No. 10, September 2009.

Nada, B., Kirton, L.G., Norma-Rashid, Y., Cheng, S., Shahlinney, L. & Phon, C.K. (2012) Monitoring the fireflies of the Selangor River. In: *Mangrove and Coastal Environment of Selangor, Malaysia* (Ed. by Sasekumar, A. & Chong, V.C.), pp. 153–162. Institute of Ocean and Earth Sciences Monograph Series 14. Kuala Lumpur: University of Malaya.

Nada, B., Kirton, L.G., Norma-Rashid, Y. & Khoo, V. (2009b) Conservation efforts for the synchronous fireflies of the Selangor River in Malaysia. In: *Diversity and Conservation of Fireflies. Proceedings of the International Symposium on Diversity and Conservation of Fireflies, Queen Sirikit Botanic Garden, Chiang Mai, Thailand, 26–30 August 2008* (Ed. by Napompeth, B.), pp. 160–171. Chiang Mai: Queen Sirikit Botanic Garden.

Nallakumar, K.A. (2002) *Management and Conservation of Fireflies in Peninsular Malaysia.* Ph.D. Thesis. University of London, Imperial College of Science, Technology and Medicine. 216 pp.

Wong, C.H. (2009) Firefly watching and conservation involving local communities in Malaysia. In: *Diversity and Conservation of Fireflies. Proceedings of the International Symposium on Diversity and Conservation of Fireflies, Queen Sirikit Botanic Garden, Chiang Mai, Thailand, 26–30 August 2008* (Ed. by Napompeth, B.), pp. 141–148. Chiang Mai: Queen Sirikit Botanic Garden.

Lampyrid 2012. **2**:174-187

Conservation of Congregating Firefly Zones (CFZs) in Peninsular Malaysia

Choong Hay Wong* & Chin Aik Yeap

Malaysian Nature Society, JKR 641, Jalan Kelantan, Bukit Persekutuan, 50480 Kuala Lumpur, Malaysia.

**Email: conservation2@mns.org.my*

ABSTRACT Malaysia is known to support populations of congregating fireflies of the genus *Pteroptyx,* which are unique to the South-East Asian and Australo-Pacific regions. However, the current distribution of these fireflies and their colonies has remained largely undocumented. *Pteroptyx* species are found along the inter-tidal mangrove river systems. These habitats are not only important ecologically but also provide ecotourism opportunities to the local communities because of the unique flashing behaviour of these fireflies. The habitats of these fireflies are fast disappearing in Malaysia, hence there is a need to document their distribution and generate awareness of their importance amongst local government authorities and agencies responsible for managing rivers. Thus, between July 2009 and May 2010, the Malaysian Nature Society (MNS) began documenting Congregating Firefly Zones (CFZs), their land use status, threats, ecotourism potential and other characteristics. Twenty eight major CFZs were surveyed. Most of the CFZs support one to four congregating firefly species. The number of CFZs is higher on the east coast, with more than twenty different display tree species recorded. Four main threats identified were river widening for flood mitigation, agriculture (e.g. oil palm plantation establishment), aquaculture ponds (for shrimp and fish), and strong light sources from shrimp ponds, swift houses and other infrastructures. The MNS hopes to present the results of these surveys to the relevant authorities to sensitise them to the existence of these fireflies and the need to conserve and manage their habitats.

Keywords: Firefly, *Pteroptyx,* distribution, conservation, ecotourism, Peninsular Malaysia

INTRODUCTION

The tropical mangrove forests of Malaysia are known to support populations of congregating fireflies of the genus *Pteroptyx* Olivier, 1902 (Lampyridae: Luciolinae), which are unique to the South-East Asian and Australo-Pacific regions. Areas in which these fireflies congregate in large numbers for their communal displays are referred to in this paper as Congregating Firefly Zones (CFZs). The current distribution of CFZs along Malaysian rivers generally remains undocumented. However, habitat and display trees are fast disappearing in Malaysia, especially in the intertidal brackish region of rivers, in which mangrove-associates such as *Nypa, Barringtonia, Thespesia, Talipariti and Sonneratia* grow. These riverine regions are protected only by river reserve guidelines, and the enforcement of any national laws is undermined by the fact that land issues fall under State jurisdiction.

There is an urgent need to protect and conserve the unique congregating fireflies as well as the inter-tidal mangrove estuaries, freshwater swamp forest and riparian forest that forms their habitat. These forests are in imminent danger of being lost as a result of rapid and indiscriminate clearing for quick socio-economic benefits. Many are unregulated, and some are cleared illegally. Climate change may also threaten the survival of the fireflies, as decreasing rain or rising sea levels will increase salt water intrusion and wave action.

To conserve the congregating fireflies, riparian forest, peat swamps, mangroves and other coastal and marine habitats need to be protected. Such forests provide many benefits. Mangrove forests provide at least US$1.6 billion each year in ecosystem services (IUCN 2010) and potential socio-economic benefits from ecotourism (e.g. firefly watching, bird watching and mangrove tourism).

The objective of the survey carried out in this study was to document existing and potential CFZs and the threats they face, as well as to determine the current status of mangrove river habitats in Peninsular Malaysia. This would become a useful future reference in conservation work and ecotourism development. It was intended as an addition to the preliminary CFZ compilation done in 2008 (Wong 2008). In the 2008 study, only 26 CFZs were documented for three unidentified inland species of congregating fireflies. Sixteen of these CFZs had firefly watching activities. Recognisable threats to the CFZs were riverside clearance for economic gain and flood mitigation, as well as the threat of light pollution. The data compiled in this and the previous surveys will be useful for decision makers and planners in local government agencies, especially those responsible for managing rivers.

The term CFZ, as used in this paper, refers to the full extent of the habitat of firefly colonies, often located along both sides of a river or a tributary or even an artificial waterway like a canal (Ohba & Wong 2004). It stretches from the margin of the river (display trees) to the back of the mangrove or tidal flood plains (Ohba & Wong 2004; Nada & Kirton 2005; MNS 2002) where the firefly larvae and their prey, such as snails, are found (Kirton et al. 2006) and where unmated adults rest during the day.

METHODS

A preliminary list of known rivers in Peninsular Malaysia that have been said to support firefly populations was compiled from various MNS reports, an unpublished MNS Annual Firefly Survey (1999), available journals, magazines, student thesis papers, news clippings and current websites, and from personal communications.

Four zones were denoted: Northern (Perlis, Kedah, Penang and North Perak), Central (South Perak, Selangor, Negeri Sembilan and Melaka), Southern (Johor) and Eastern (Kelantan, Terengganu and Pahang). In each zone, a selection of rivers well known to be inhabited by fireflies were chosen. A selection of mangrove rivers with potential to support firefly colonies was also made, based on sightings of flash displays and interviews with villagers. The survey involved visits to sites on both listings.

Surveys took place during a 10-day window of dark nights each month, during which the moon rose after 10:00 pm or set before dusk. The surveys were carried out between 6:00 pm and 10:00 pm from July 2009 to May 2010, after the first fireflies flashed. GPS waypoints were taken using a GARMIN GPSMap 60CSx.

Daytime rapid assessment datasheets (Hashim 2009) were used to document the land use and threats, flora and fauna, and environmental conditions surveyed by boat and car. Interviews were conducted with the local residents about the area in general, firefly watching, tourism activities and conservation status. Other relevant data was also obtained.

During night surveys, representative fireflies were confirmed as belonging to the genus *Pteroptyx* by checking whether adult males had deflexed elytral apices (Ballantyne & McLean 1970). No firefly specimens were collected except during a detailed study at the Kerteh and Cherating Rivers. Information recorded on night datasheets (Hashim 2009) included the extent of firefly colonies, the occurrence of synchronous or non synchronous flashing, flash rates and colour, locations of larvae, and the presence of human activities and light pollution. The display trees were identified where possible. Trees with a high density of fireflies were recorded, either as individual trees or as a stretch of trees. A table of CFZs in Peninsular Malaysia, compiled according to States, is presented together with a map.

Not all rivers had their whole CFZs surveyed, as weather and navigational difficulties sometimes prevented a complete survey. Some of the data was based on information provided by boatmen. Big rivers with many tributaries, like the Perak and Bernam River, required a few visits.

RESULTS

Congregating fireflies in Malaysia

Currently, there are 8 species of *Pteroptyx* recorded from Malaysia (Ballantyne 2001). Six of them are confirmed to congregate on river-estuary trees (*Pteroptyx bearni, P. gelasina, P. malaccae, P. similis, P. tener* and *P. valida*) and two are unverified (*P.*

asymmetria and *P. decolor*) (see also Ohba & Wong 2012).

Congregating Firefly Zones (CFZs)

From physical and literature surveys, 58 CFZs were determined, with more anticipated to be discovered. Of these, 28 CFZs were physically surveyed. CFZs were identified in nine states: Kedah (2), Penang (1), Perak (7), Selangor (2), Melaka/ Negeri Sembilan (1), Johor (2), Pahang (10), and Terengganu (3). Remaining mangroves not covered were in Terengganu (north coast), Johor (north-east and west coast), Perak (Matang to Pasir Hitam area) and major islands (Langkawi, Labuan and Tioman). A compilation of the CFZs in Peninsular Malaysia according to States is given in Table 1, and Figure 1 shows the location of each CFZ on a map.

Ten CFZ are located in Protected Areas like Ramsar Sites, State Parks and forest reserves. Malaysia's only protected area specifically for fireflies is part of the Selangor River, protected under Government of Selangor Gazette 63(9), no. 1188, effective 8 April 2010.

Firefly watching activities in the CFZs were managed by local community groups, the government, private individuals, private companies, or partnerships (Table 2). There were also CFZs in which we were not able to determine whether firefly watching activities were carried out, and potential sites in which firefly watching could be promoted. Awareness, however, is still low among members of the local communities living in the vicinity of CFZs.

Major threats to CFZs

The major problems faced by the CFZs pertained mainly to unsustainable land use and economic activities. Primary threats identified were clearing of river reserves, strong night light sources, pollution, unregulated tourism, exploitation of natural resources, bank erosion and, potentially, climate change. Four main activities were identified: river widening (for flood mitigation), agriculture (oil palm), aquaculture (shrimp and fish ponds), sand mining (illegal and legal). Some of these activities were also responsible for strong night light sources (e.g., shrimp ponds, swift houses and infrastructures, including tourism-related). There were also unregulated and illegal activities such as land clearing, timber harvesting and wildlife poaching being carried out at some CFZs.

Other commercial activities in and around the CFZs produce waste (sullage, sewage, rubbish and factory effluents) and chemical pollutants (pesticides and fertilisers) that contaminate water and threaten firefly larvae and their prey. River bank erosion from boat wakes, and increased salt water intrusion as a result of water extraction, weirs or climate change also affect fireflies and the riverine vegetation. These threats are prevalent along the more developed west coast of the Peninsula, but are beginning to surface in the east coast as the new East Coast Economic Region

	State	River / Area	CFZ or site coordinates	Congregating firefly species	CFZ protected status	Firefly watching	Sources of information
1	Kelantan	Tumpat Lagoon	N6°12', E102°13'	Data deficient			MNS firefly survey in 1999 (unpublished)
2	Kelantan	Semerak Lagoon (s)	N5°51', E102°30'	*Pteroptyx malaccae*	Mangrove reserve		Wahizatul-Afzan & Zainuddin 2005; Zainuddin 2006
3	Terengganu	Setiu-Chalok Lagoon (s)	N5°35', E102°49'	Rain, no survey was done; observed a tree at Chalok	Proposed Setiu Wetlands State Park		Pers. comm. with villagers
4	Terengganu	Sungai Dungun trib.	N4°46', E103°24'	Data deficient			Pers. comms. with villagers
5	Terengganu	Paka	N4°38', E103°24'	Data deficient			Pers. comms. with villagers
6	Terengganu*	Kerteh (s), (n)	N4°31', E103°26'; N4°34', E103°23'	*Pteroptyx tener*, *P. bearni* and non-synchronous species			Ohba & Wong 2012
7	Terengganu*	Chukai and trib. (s), (n)	N4°18', E103°22'; N4°15', E103°24'	*Pteroptyx tener*, *P. bearni* and non-synchronous species		yes	
8	Terengganu	Kemaman	N4°14', E103°25'	Data deficient			Pers. comms. with villagers
9	Pahang*	Cherating (n)	N4°48', E103°23'; N4°7', E103°23'	*Pteroptyx bearni*	Permanent forest reserve	yes	Ohba & Wong 2012
10	Pahang*	Balok (n)	N3°56', E103°21'; N3°56', E103°22'	One non-synchronous *Pteroptyx* sp.			
11	Pahang	Kuantan	N3°48', E103°19'	Data deficient		yes	Pers. comm. with villager; tours websites-http://www.kuantanboating.mysms123.com/firefly-wathcing.html; http://malaysiahotelnews.blogspot.com/2007/03/light-dims-for-fireflies-along-sungai.html
12	Pahang*	Belat (trib. of Kuantan) (s), (n)	N3°45', E103°16' -N3°45', E103°15'	*Pteroptyx tener* (more on nipah); one non-synchronous species		yes	
13	Pahang*	(1) Pulau Bijeh (island); (2) Pulau Miseh (island); (3) river mouth of Pahang	(1) N3°30', E103°27'; N3°31', E103°27' (2) N3°31', E103°26'; N3°30', E103°26' (3) N3°31', E103°26'; N3°30', E103°25'	One synchronous *Pteroptyx* sp. and one non-synchronous *Pteroptyx* sp.			
14	Pahang*	Bebar (n)	N3°08', E103°26'; N3°09', E103°25'	One non-synchronous *Pteroptyx* sp.		yes	

Table 1 The CFZs in Peninsular Malaysia with non synchronous and synchronous fireflies; n = non synchronous; s = synchronous; trib. = tributary/ tributaries); * indicates the site was physically surveyed.

	State	River / Area	CFZ or site coordinates	Congregating firefly species	CFZ protected status	Firefly watching	Sources of information
15	Pahang*	Merchong (s), (n)	N3°01', E103°24'; N3°0', E103°25'	Two non-synchronous *Pteroptyx* sp.; one synchronous *Pteroptyx* sp.		yes	
16	Pahang*	Rompin (s), (n)	N2°51', E103°25'; N2°53', E103°23'	Two non-synchronous *Pteroptyx* sp.; one synchronous *Pteroptyx* sp.		yes	
17	Pahang*	Pontian (s), (n)	N2°45', E103°27'; N2°45', E103°30'	Two non-synchronous *Pteroptyx* sp.; one synchronous *Pteroptyx* sp.			
18	Pahang	Mentawak, Tioman Island	N2°47', E104°12'	Data deficient			Pers. comm. with villager
19	Pahang	Barok, Tioman Island	N2°47', E104°12'	Data deficient			Pers. comm. with villager
20	Pahang*	Anak Endau (trib. of Endau) (n)	N2°41', E103°35'; N2°39', E103°36'	Two non-synchronous *Pteroptyx* sp.			
21	Pahang-Johor border*	Endau (s), (n)	N2°39', E103°35'; N2°38', E103°36'	Two non-synchronous *Pteroptyx* sp.; *P. tener*			Ballantyne 2001
22	Johor	Sedili Besar	N1°56', E104°06'	Data deficient		yes	http://www.anglochineseprimary.net/2010corres/pdf/ECOCAMP_3D2N_ACSP%20-%202010-05-28.pdf
23	Johor	Sedili Kecil, Sg Kangkar	N1°49', E104°08'	Reduced due to flood mitigation			Road visit only ; pers.comms with villagers and NGO staff
24	Johor	Mawai (s)	N1°52', E103°57'	*Pteroptyx tener*			Ballantyne 2001
25	Johor*	Johor and trib. (Nam Heng Estate) (s), (n)	N1°43', E103°54'; N1°40', E103°56'	*Pteroptyx malaccae; P. tener;* one non-synchronous *Pteroptyx* sp.		yes	Ballantyne 2001
26	Johor*	Lebam & trib. (s), (n)	N1°33', E104°10'; N1°34', E104°11'	Non-synchronous firefly (possibly *Pteroptyx bearni)*	Forest Reserve	yes	
27	Johor	Pulai	N1°20', E103°32'	Data deficient	Forest Reserve		Pers. comm. with biologist
28	Johor	Tanjung Piai (bunds) (n)	N1°16', E103°30'	One non-synchronous species	Ramsar Site/ State Park	yes	Pers. comms. with State Park staff
29	Johor	Pulau Kukup and Kukup (canal)	N1°19', E103°26'	Data deficient	Ramsar Site/ State Park		Pers. comms. with State Park staff
30	Johor	Benut (s)	N1°38', E103°15'	*Pteroptyx malaccae; P. tener*			Pers. comms with State Park staff; Nallakumar 2002; Bassot & Polunin 1968; Ballantyne & McLean 1970; Ballantyne 2001

Table 1 Continued.

	State	River / Area	CFZ or site coordinates	Congregating firefly species	CFZ protected status	Firefly watching	Sources of information
31	Johor	Muar	N2°02', E102°33'	Data deficient		yes	Media website-http://www1.nst.com.my/Current_News/JohorBuzz/Wednesday/Stories/20091014094020/Article/
32	Johor	Kesang	N2°07', E102°30'	Data deficient			Pers. comms with villager
33	Johor	Kupia Labong, trib. of Endau (s)	N2°35', E103°37'	*Pteroptyx tener*			Ballantyne 2001
34	Melaka-Negeri Sembilan border*	Sungai Linggi-Rembau (s), (n)	N2°26', E102°03'; N2°26', E102°03'	*Pteroptyx tener*; *Pteroptyx* sp.		yes	Jusoh et al. 2010a
35	Selangor	Sepang Kecil & Sepang Besar	N2°36', E101°41'	Reduced			Pers. comms. with MNS sepang education interpretive center staff
36	Selangor	Langat (n)	N2°50', E101°25'	*Pteroptyx assymetria* (fragmented)			Pers. comm. with villager; observed from land; Ballantyne 2001
37	Selangor	Rambai, Carey Island	N2°51', E101°17'	Data deficient			Pers. comm. with villager
38	Selangor	Buloh	N3°15', E101°18'	Data deficient			From media
39	Selangor*	Selangor (s)	N3°22', E101°19'; N3°21', E101°16'	*Pteroptyx tener; P. malaccae*	Protection Zone under LUAS	yes	Ballantyne 2001; Ohba & Wong 2012
40	Selangor*	Kuala Selangor Nature Park (n)	N3°20', E101°14'	*Pteroptyx valida; Colophotia* sp.	Town Council Park		MNS firefly survey in 1999 (unpublished); Ballantyne 2001
41	Selangor-Perak Border*	Bernam (s), (n)	N3°46', E100°57'; N3°51', E100°56'	One synchronous and one non-synchronous *Pteroptyx* sp.		yes	
42	Perak*	Bidor (a trib. of Perak) (s), (n)	N4°01', E101°01'; N4°00', E101°03'	One synchronous and one non-synchronous *Pteroptyx* sp.			
43	Perak*	Dedap (a trib. of Perak) (s), (n)	N3°58', E100°53'; N3°57', E100°52'	One synchronous and one non-synchronous *Pteroptyx* sp.			
44	Perak	Sumum and Rebana (trib. of Perak)	N3°54', E100°51'	Data deficient			MNS firefly survey in 1999 (unpublished)
45	Perak*	Perak (s), (n)	(1) N4°03', E101°01'; N4°03', E101°01'; (2) N3°58', E100°52'; N3°58', E100°53'; (3) N4°0', E100°48'; N4°0', E100°46'; *N3°59', E100°45'	(1) One synchronous & one non-synchronous *Pteroptyx* sp.; (2) One synchronous & one non-synchronous *Pteroptyx* sp. (3) one synchronous & two non-synchronous *Pteroptyx* sp. (*) unknown			Surveyed (visit in three separate occasions) a few fireflies were flying individually among trees and could not be identified with certainty

Table 1 Continued.

	State	River / Area	CFZ or site coordinates	Congregating firefly species	CFZ protected status	Firefly watching	Sources of information
46	Perak*	Sitiawan and trib. (n)	N4°14', E100°41'	Only two individuals spotted: possibly *Pteroptyx assymetria* (disturbed riverbank); *P. valida*			Ballantyne 2001
47	Perak	Pasir Hitam	N4°38', E100° 37'	Data deficient			Tours website-http://outskirtout-reach.org/blog/fire-fly-kuala-selangor-kl-kuala-lumpur-tour-ecotourism-tourist/
48	Perak	Terong	N4°40', E100°39'	Data deficient			Tours website-http://outskirtout-reach.org/blog/fire-fly-kuala-selangor-kl-kuala-lumpur-tour-ecotourism-tourist/
49	Perak*	Sepetang (s), (n)	N4° 54', E100°39'; N4°52', E100°38'	*Pteroptyx tener*; one unidentified non-synchronous firefly	Forest Reserve (partial)	yes	Salmi et al. 2009; Zaidi et al. 2005; Jusoh et al. 2010b
50	Perak*	Bund Choo Kao, Teluk Rubiah, Kuala Gula (n)	N4°55', E100°27'; N4°55', E100°27'	One non-synchronous congre-gating firefly		yes	
51	Perak*	Kurau (s), (n)	N5°00', E100°26'; N4°59', E100°27'	One synchronous and one non-synchronous *Pteroptyx* sp.			
52	Penang *	Batu Maung mangroves (n)	N5°16', E100°14'	*Pteroptyx assymetria*			Ballantyne 2001
53	Penang *	Balik Pulau mangroves	N5°23', E100°11'	Data deficient			Pers. comm. with teacher
54	Penang *	Krian, Byram Estate (s)	N5°09', E100°30'; N5°09', E100°26'	*Pteroptyx tener*		yes	Pot 2009
55	Kedah*	Merbok (n)	N5°42', E100°29'; N5°41', E100°29'	Two non-synchronous *Pteroptyx* sp.			
56	Kedah*	(1) Lallang and (2) Bedong (tribs. of Merbok) (n)	(1) N5°41', E100°30'; N5°41', E100°30' (2) N5°41', E100°30'; N5°42', E100°30'	(1) Two non-synchronous *Pteroptyx* sp. (2) One non-synchronous *Pteroptyx* sp.			
57	Kedah	Kubang Badak , Pulau Langkawi	N6°24', E99°43'	None spotted after 3 days of heavy rain	Permanent Forest Reserve		Pers. comms. with MNS member and tour operator
58	Perlis	Sungai Perlis	N6°24', E100°07'	Data deficient			Pers. comm. with MNS member

Table 1 Continued.

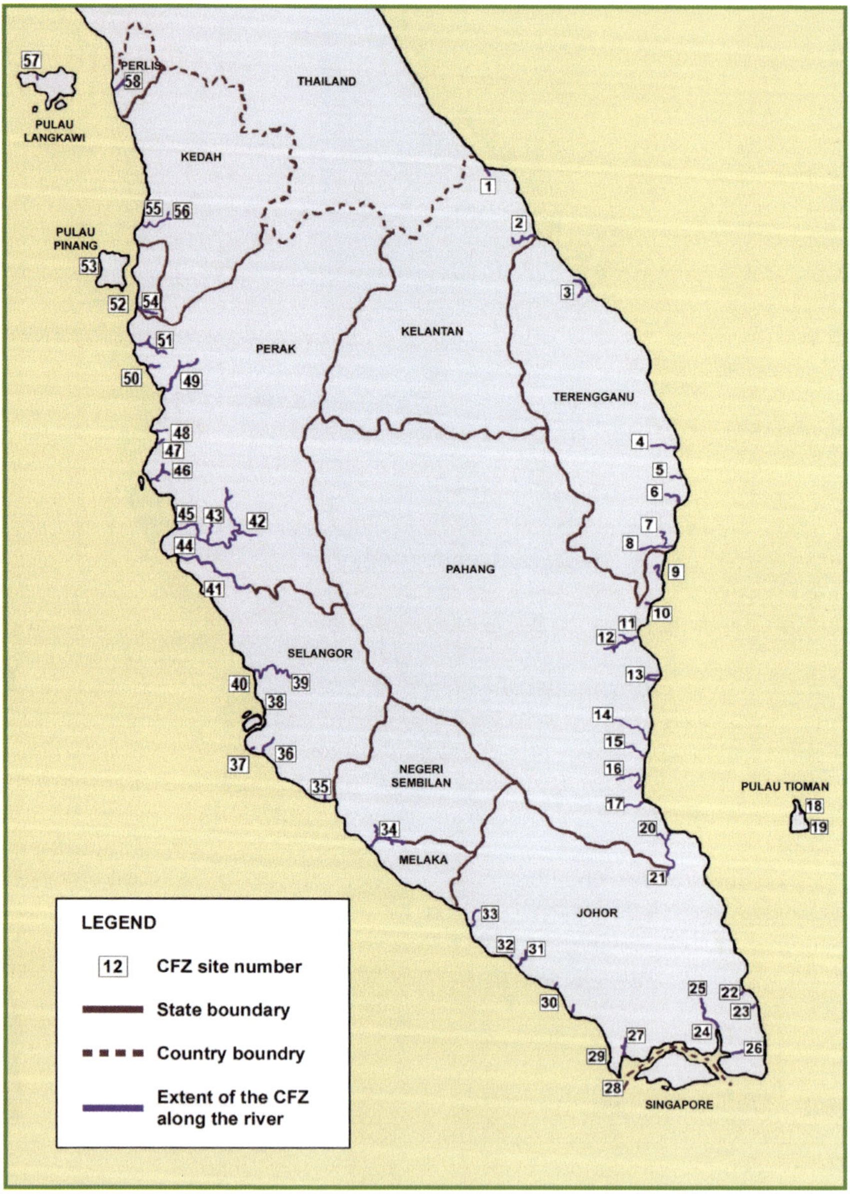

Figure 1 The 58 CFZs in Peninsular Malaysia (adapted from the Conservation of Congregating Firefly Zones (CFZs) in Malaysia poster by MNS in 2010).

Firefly watching sites	A	B	C	D	E	F(i)	F (ii)	F (iii)	G	H	Total CFZs
CFZs surveyed	23	5	0	1	5	1	1	1	1	0	28
Other CFZs	5	1	0	0	4	1	1	0	0	46	30

A. Potential firefly watching area: Walk in visitors or based on word of mouth; taken by local boatmen based on an agreed price; few visitors in a year or every few years. **B.** Local community group (LCG): Villagers formed a group to facilitate firefly watching, either under the Fishermen's Association or the Village Welfare and Safety Committee (JKKK), or as part of a homestay programme; some work full time; most earn a supplementary income. **C.** NGOs: Non-governmental organisations or non-profit organisations. **D.** Government: Either a local government or an institutional body. **E.** Private enterprise or companies with or without a nature guide license, or a group of full-time local boatmen and guides or non-local tour operators who sometimes send customers to the locals or LCGs. **F.** Partnerships between the government and (i) a LCG, (ii) an NGO, (iii) an enterprise (the government helps directly or indirectly in infrastructure construction (eg. jetties and in promotng tourism, and the area is owned by the government; some are linked with hotels, e.g. Summerset Resort in Rompin, or with other organisations that serve the tourist). **G.** LCG+NGO partnership: Capacity building and formation of a LCG by an NGO. **H.** Unknown: Rivers that were not visited, so no information is available.

Table 2 Firefly watching sites in Peninsular Malaysia.

(ECER) development takes place.

Habitats of the fireflies

Most of the west coast rivers appear to be more disturbed and silted, with fewer major CFZs compared to rivers on the east coast. Although having fewer CFZs, the firefly colonies observed in some of the west coast rivers are relatively dense. Many colonies were fragmented over single trees or stretches of trees. A single species of *Pteroptyx* was frequently found in west coast sites.

In contrast, the east coast rivers appeared to be relatively undisturbed. Here *Pteroptyx* were observed to congregate and display on a wide variety of riparian and mangrove trees (especially *Sonneratia caseolaris)*, low shrubs, climbers, vines, a species of palm and even an exotic tree species (*Acacia mangium)*. A greater variety of display trees were surveyed along the Kerteh River, Terengganu in 2009, with as many as twenty seven species being used by the fireflies, especially in the brackish region. It appeared to be common to have more than one species of *Pteroptyx* occurring at one riverine site (c.f. Ohba & Wong 2012).

It was observed that the *Pteroptyx* had a preference to display on smaller, young leaves such as we found on *S. caseolaris* in the Kurau, Bernam and Selangor Rivers after an attack by mealy bugs. Small-leafed trees like terong tikus along the Kerteh River were also observed to be preferred display trees. It was also observed that the firefly would display on old, tall *S. caseolaris* trees with sparse leaves.

Only one palm tree, *Nypa fruticans,* was observed to be used as a display tree by congregating fireflies. The use of this palm as a display tree was observed in at least three rivers, two on the east coast. The *Nypa* palm was observed to be one of the more preferred display trees.

DISCUSSION

Most of the west coast rivers were more disturbed than the rivers on the east coast, but the patches of firefly colonies observed at some of the west coast rivers were formed by relatively dense aggregations of fireflies. One factor contributing to this could be a more open environment. As Mahadimenakbar et al. (2007) have postulated for *Pteroptyx* in Sabah, proximity to grassland allows the fireflies a clear view for flash communication. Tall grass also serves as an important refuge for the larvae of *Pteroptyx valida* during the daytime, and a place where female fireflies can lay their eggs (Ohba & Sim 1994).

The east coast rivers appeared to have two species of firefly that inhabited different vegetation zones of the river, sometimes with an overlap. The segregation of habitat may increase over time as the river vegetation undergoes natural change, and may eventually totally isolate the different species. The isolation of the two species shows their affinity for different habitats and vegetation. It appears that the non-synchronous firefly species has a preference for saline regions that generally occur downstream of *Nypa*, while the synchronous species prefers the more upstream brackish region. It may be possible to predict the firefly species that will occur in a particular region of a river based on the tree species present.

Sonneratia caseolaris appeared to be a generally reliable indicator of the presence of fireflies, to which most villagers could testify. This was especially evident from the Kerteh River. The affinity of *Pteroptyx* fireflies for the younger leaves of *S. caseolaris* during the night could be due to territorial behaviour, the fireflies preferring to be on isolated leaves when displaying. However, during the day, the fireflies were observed to share single leaves.

In some west coast rivers, like the Krian River and Perak River, much of the riverbanks has been cleared and bunded a long time ago. Display trees in these areas comprised small groups or solitary trees, or were amongst other vegetation like scrub, grassland and *Nypa* palms. These trees had good displays of fireflies, possibly because remnant colonies were breeding nearby.

Firefly ecotourism and conservation

Firefly watching in Malaysia is a year-round activity, although fewer fireflies are seen during the north-east monsoon from October to January. A good CFZ will be able to attract a lot of tourists. Involvement of the local community in firefly watching activities appears to be increasing and may play a role in preserving the habitat of the fireflies because of the perceived economic benefits. The local community can benefit in two ways. Firstly, by conserving natural habitats for fireflies, their livelihood, which depends on natural resources, will be better ensured. Secondly, they can opt for either full-time employment in the ecotourism industry or just supplement their income by working as guides and boat operators. However, if the economic activities of the local community move away from ecotourism to agriculture, aquaculture or other intensive land-use related activities, this may pose a threat to conservation of the fireflies, especially if land is used within the river reserve. There is a need for a general framework for a firefly ecotourism and conservation management plan that covers existing and potential firefly ecotourism sites. This would help in the protection of the sites and businesses (Wong 2008).

Protecting CFZs

Most of the major threats to the CFZs are a result of a lack of enforcement and monitoring of the river reserves, and poor legal status for protection. There is also a lack of basic research on rehabilitation and conservation of firefly habitats, and on climate change induced by human activities, which may also indirectly affect CFZs.

Recommended solutions to minimise or eliminate the threats may include a River Reserve Act for better legal protection status. New and less damaging flood mitigation methods are needed, as well as planting on bunds. Other needs are awareness and education of boatmen and guides, speed limits for motor boats, reduction of light pollution (blocking, lowering or redirecting light sources), reduction of pesticide usage, and proper waste disposal systems.

CONCLUSION

The compilation of CFZs undertaken in this study will serve as a baseline reference for conservation, awareness and monitoring work. It can be regularly updated and used as a tool for resource managers, land use planners, decision makers and policy makers, and is also a valuable reference for the public. Urgent formulation of zonations is needed, as well as conservation and management plans for the CFZs. Future surveys will help update and add to the compilation of CFZs.

ACKNOWLEDGEMENTS

This project would not have been possible without support from all our colleagues, MNS members, and friends. Special thanks to Nor Rasidah Hashim and Wan Faridah Akmal Wan Jusoh from Universiti Putra Malaysia for the design of the data sheets, and to Chey Vun Khen, the late Ivan Polunin, Lesley Ballantyne, Loh Chi Leong, Mahadimenakbar M. Dawood, Ohba Nobuyoshi, and Mohd Rozaimi Jamaluddin for their kind advice and encouragement and for the provision of information and materials used in collating this paper. We also greatly thank all the experienced and knowledgeable boatmen who assisted us. This work was undertaken with the support of the Mohamed Bin Zayed Species Conservation Fund, project no. 0925338.

REFERENCES

BALLANTYNE L.A. & MCLEAN M.R. (1970) Revisional studies on the firefly genus *Pteroptyx* Olivier (Coleoptera: Lampyridae: Luciolinae: Luciolini). *Transactions of the American Entomological Society* **96** 223-305.

BALLANTYNE L.A. (2001) The bent winged fireflies of Cambodia, Indonesia, Malaysia, Philippines and Thailand (Coleoptera: Lampyridae: Luciolinae: Luciolini), *Pteroptyx* spp. of the Polunin Collection. *Serangga* **6** 51-95.

BASSOT, J.M. & POLUNIN, I. (1968) Synchronously flashing fireflies in the Malay Peninsula. *Science Report of the Yokosuka City Museum* **13** 18-23.

HASHIM, N.R. (2009) *Manual for field survey & SWOT analysis of congregating fireflies in Malaysia. Based on the experiences of firefly researchers at Universiti Pertanian Malaysia & Malaysian Nature Society.* Universiti Pertanian Malaysia, 20 pp.

IUCN. (2010) *Mangrove forests in worldwide decline.* http://iucn.org/about/work/programmes/species/?5025/Mangroves-forests-in-worldwide-decline....4/17/2010.

JUSOH, W.F.A.W., HASHIM, N.R. & IBRAHIM, Z.Z. (2010a) Distribution and abundance of *Pteroptyx* fireflies in Rembau-Linggi Estuary, Peninsular Malaysia. *Environment Asia* **3** 56-60.

JUSOH, W.F.A.W., HASHIM, N.R. & IBRAHIM, Z.Z. (2010b) Firefly distribution and abundance on mangrove assemblages in Sepetang estuary, Peninsular Malaysia. *Wetlands Ecology and Management* **18** 367-373.

KIRTON, L.G., NADA, B., TAN, S.A., ANG, L.H., TANG, L.K., HUI, T.F. & HO, W.H. (2006) The Kampung Kuantan Firefly Project: A preliminary assessment of the habitat requirements of *Pteroptyx tener* (Coleoptera: Lampyridae). In: *Highlights of FRIM's Non-IRPA Projects* (Ed. by Ho, Y.F. & Sarifah, K.A.), pp. 119-121. Selangor: Forest Research Institute Malaysia.

NALLAKUMAR, K.A. (2002) *Management and Conservation of Fireflies in Peninsular Malaysia.* Ph.D. Thesis. University of London, Imperial College of Science, Technology and Medicine, London. 216 pp.

MAHADIMENAKBAR, M.D., HAIRUL, H.M. & MAZIDI, A.G. (2007) The distribution and phenology of *Pteroptyx* fireflies (Coleoptera: Lampyridae) along Garama River, Sabah, Malaysia. *Journal of Tropical Biology and Conservation* **3** 1-9.

MNS (MALAYSIAN NATURE SOCIETY). (2002) *Firefly Studies in Kuala Selangor District.* Final Report submitted to the Selangor State Government. Kuala Lumpur: Malaysian Nature Society.

NADA, B. & KIRTON, L.G. (2005) Let there be light, firefly light. In: *FRIM in Focus*, pp.

8-9, April-June 2005.

OHBA, N. & SIM, S.H. (1994) The morphology, behavior and life cycle of *Pteroptyx valida* (Coleoptera: Lampyridae) in Singapore. *Science Report of the Yokosuka City Museum* **42** 1-11.

OHBA, N. & WONG, C.H. (2004) External morphology and ecological study of the firefly *Pteroptyx tener* at Kampong Kuantan, Selangor, Malaysia. *Science Report of the Yokosuka City Museum* **51** 1-33.

OHBA, N. & WONG, C.H. (2012) Observations on the external morphology and behaviour of *Pteroptyx tener* and *Pteropytx bearni* in Malaysia. In: *Firefly Conservation: From Science to Practice. Proceedings of the Second International Firefly Symposium, Selangor, Malaysia, 2-5 August 2010* (Ed. by Kirton, L.G., Lim, G.T. & Day, J.C.). *Lampyrid* **2** 80-86.

POT, S.S. (2009) *Fireflies as a Nature Tourism Product in Nibong Tebal, Penang.* B.Sc. Thesis. Universiti Malaysia Sabah.

SALMI, N., LATIP, A., YAHAYA, N.N.F. & RAHIM, A.A. (2009) Fireflies distribution in relation to human development in the Sepetang River catchments. In: *Proceedings of the Research Seminar on Salt Water Intrusion in Kuala Sepetang, Malaysia, Taiping, Perak, Malaysia, 4 Jun 2009* (Ed. Shaaban, A.J.), pp. 41-55. Malaysia: National Hydraulics Research Institute Malaysia (NAHRIM).

WAHIZATUL-AFZAN, A. & ZAINUDDIN, A. (2005) *Notes on Pteroptyx malaccae (Coleoptera: Lampyridae) with Their Host Plant Preference in Mangrove Forest of Tok Bali, Kelantan.* Unpublished report. Department of Biological Sciences, Faculty of Science and Technology, Kolej Universiti Sains dan Teknologi Malaysia (KUSTEM).

WONG, C.H. (2008) Firefly watching and conservation involving local communities in Malaysia. In: *Diversity and Conservation of Firelies. Proceedings of the International Symposium on Diversity and Conservation of Fireflies, Queen Sirikit Botanic Garden, Chiang Mai, Thailand, 26-30 August 2008* (Ed. by Napompeth, B.), pp. 141-148. Chiang Mai: Queen Sirikit Botanic Gardens.

ZAIDI, M.I., AZMAN, S. & WONG, C.Y. (2005) Synchronous fireflies on berembang trees along Sungai Sepetang, Taiping, Perak. In: *Sustainable Management of Matang Mangroves: 100 years and Beyond. Proceedings of the national Conference. Forest Biodiversity Series 4, Ipoh, Perak, Malaysia, 5-8 October 2004* (Ed. by Shaharuddin, M.I., Azahar, M., Razani, U., Kamaruzaman, A.B., lim, K.l., Suhaili, R., Jalil, M.D. & latiff, A.), pp. 282-287. Kuala Lumpur: Forest Department of Peninsular Malaysia.

ZAINUDDIN, A. (2006) *Composition and Abundance of Firefly Community at Tok Bali Mangrove Forest and Kampung Tandak, Kelantan.* B.Sc. Thesis. Kolej Universiti Sains dan Teknologi Malaysia (KUSTEM).

Lampyrid 2012. **2**:188-191

Pirilampos (The Fireflies) — An educational workshop on fireflies combining the arts and sciences

Maria Mendonça*, José Fidalgo & Luís Vasconcelos

Associação Portuguesa dos Pequenos Investigadores, Caminho Poço do Monte, nº2, 4705 - 671 - Braga, Portugal.
**Email: cmilhais@gmail.com*

ABSTRACT Pirilampos was a five-day workshop about fireflies held in the Museu dos Biscaínhos, Braga, Portugal. It was aimed at children from 5 to 12 years old. The objective of the workshop was to introduce fireflies to children in the perspective of science, plastic arts and dance. The innovative collaboration between instructors in the sciences and arts enabled workshop participants to develop multiple skills. The art school Arte Total and the Associação Portuguesa dos Pequenos Investigadores, both based in Braga, Portugal, were responsible for the art and science components, respectively.

Keywords: Fireflies, child education, workshop, arts, sciences

INTRODUCTION

Pirilampos was a five-day, firefly-themed children's workshop held in the Museu Dos Biscaínhos, Portugal. The workshop was jointly organised by the Associação Portuguesa dos Pequenos Investigadores (APPI) and the art school Arte Total. The objective of the workshop was to introduce fireflies to children using the perspectives of science, plastic arts and dance.

The Portuguese Association of Young Researchers (APPI) is a non-profit association established for the main purpose of educating young people in science. Its members are all professionals from diverse fields who work for the association as volunteers. One of the activities of APPI is the organisation of educational workshops on subjects in science, such as electricity, water, air, colours and fireflies, through the approaches of artistic and body expression. Most workshops are a week long. To facilitate artistic and body expression, APPI has as a partner a school named Arte Total. It is an art school that aims to spread and develop dance, theatre and plastic arts. Both APPI and Arte Total schools are located in Braga, a city in northern Portugal.

Figure 1 The venue of the firefly workshop, Museu dos Biscaínhos.

Another of APPI's partners is the Museu dos Biscaínhos. This museum consists of a seventeenth century manor house built of granite, surrounded by extensive gardens (Fig. 1) that still maintain the original design. The opportunity to work in this historic and beautiful setting is a great advantage for the workshops, which contributes to the children's successful learning.

This paper describes an innovative approach of integrating science and the arts to teach, using fireflies as the focal subject. Activities for each workshop component are outlined. The impact of the workshop on attitudes towards fireflies and firefly conservation are discussed.

WORKSHOP PROGRAMME

The workshop on fireflies was a week long and involved three types of activities focusing on science, plastic arts and dance. Every day there was work involving the three components. The work was developed from the scientific component, around one initial question (e.g., "What are fireflies?" "What do they eat?" "Where do they live?"), which drew the learner into the other remaining components.

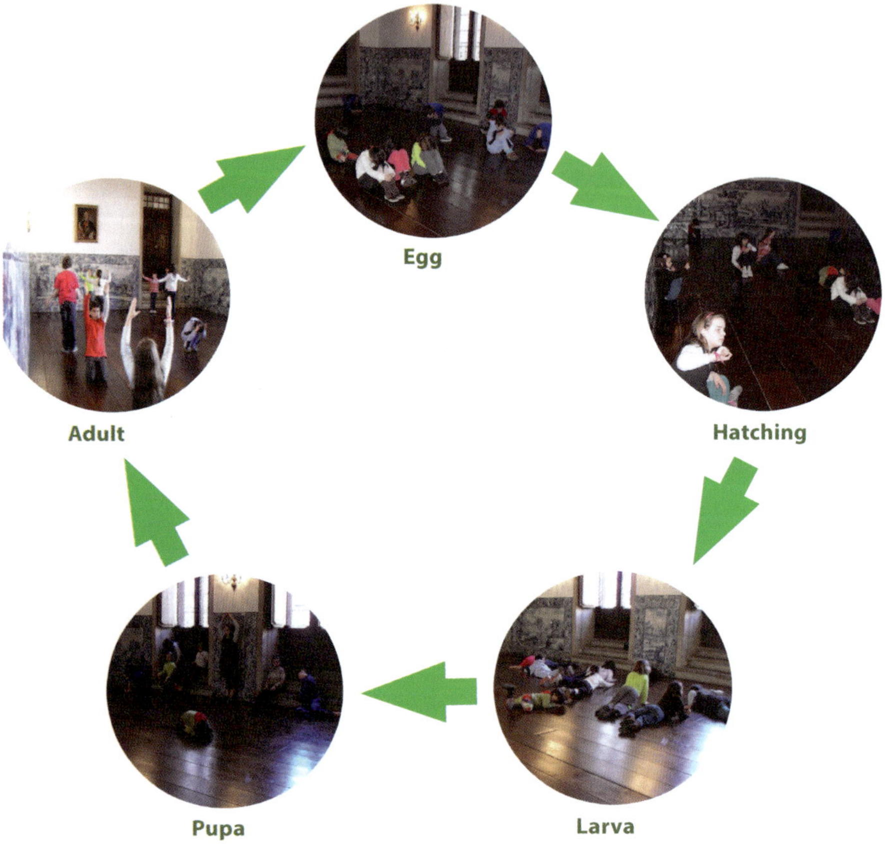

Figure 2 Dance performance depicting the firefly life cycle.

In the science component, the fireflies were viewed from the perspective of biology, physics and chemistry. The biology perspective related to the life cycle of the firefly and the environmental factors required for its survival. With respect to physics, there was discussion about types of light sources (incandescent, fluorescent, LEDs, warm light and cool light) to help arrive at the conclusion that the light emitted by fireflies is ideal. In relation to chemistry, the young participants got to know and work with luminol, experimenting with its reaction with oxygenated water in a plastic bottle, to obtain a glow similar to that of fireflies.

The plastic arts component had children draw and craft fireflies using paper and cardboard, before and after being shown pictures of real fireflies. In some of the creations, LEDs were used to emit light. Some of the crafted fireflies were used to make jewellery such as necklaces, earrings and hair clips.

In the dance component, children were encouraged to artistically act out the stages of the firefly life cycle (Fig. 2). The dance was accompanied by a soundtrack from the movie "The Lion King," entitled "The Circle of Life."

In the afternoon of the last day of the workshop, children showed their parents what they had learnt about fireflies and the natural conditions necessary for their survival. They presented and explained the experiments they had done in science and guided their parents through an exhibition of crafts made in the plastic arts.

In Portugal, one of the few places that support colonies of fireflies is the "Parque Biológico de Gaia" (Gaia Biological Park) where the First Firefly Network Meeting was held in 2007. This park is located in Vila Nova de Gaia, a city situated 50 km south of Braga. The park conducts guided visits in June every year when Portugal's few firefly species appear in significant numbers.

It is important to note that most children begin the workshop without ever having seen even a drawing or photograph of a firefly. This is obvious when they try to draw a firefly for the first time. They draw very bizarre little animals that have nothing to do with reality.

During the biology component of the workshop, participants were made aware of the existence of fireflies at the Gaia Biological Park and were invited along with their families to view the fireflies after the workshop. Almost all the children that attended the workshop have visited the park, not only in that year but also in the subsequent years, taking their relatives and friends, who end up sharing the enthusiasm of the students' new knowledge of these small and wonderful creatures. These new enthusiasts have effectively contributed to the dissemination of the wonder of fireflies.

It has been good to know that children who attended these workshops years ago still remember, with great enthusiasm, details of what they experienced. Perhaps this is why the project has been so successful, with a great demand for these workshops, and enrolment filling quickly every year.

CONCLUSION

The learning technique used in this new approach has the merit of developing skills in plastic arts, drama and body expression, providing a sense of aesthetics and rhythm. In addition, it has raised awareness of the importance of taking care of nature and of the vulnerability of fireflies. The workshop has turned out to be very important in the spread of information about fireflies in Portugal, and in making people aware of measures that have to be taken to stop their disappearance. Even the high initial expectations of this program were exceeded by its success.

Lampyrid 2012. **2**:192-196

Abstracts of Other Papers Presented

The evolution of bioluminescence in Coleoptera

MARC A. BRANHAM[1*] & JESSICA LIGHT[2]

[1]*Department of Entomology and Nematology, University of Florida, Gainesville, FL. USA.*
[2]*Department of Wildlife and Fisheries Sciences, Texas A+M University, College Station, TX. USA.*
**Email: marcbran@ufl.edu*

We employ modern phylogenetic techniques to identify the position and date the origins of bioluminescence in the insect order Coleoptera. Our phylogenetic analysis was composed of 137 taxa selected to represent the diversity within Elateriformia, and sequence data was obtained from three genes, COI, 16S and 18S. We used the resulting phylogeny to estimate divergence times and calibrated the analysis using fossils from the families Lampyridae and Elateridae. We discuss the relative age of the bioluminescent lineages within Coleoptera and its relationship to morphological and behavioural complexity associated with these bioluminescent signal systems.

Keywords: Bioluminescence, evolution, phylogeny, beetles, Coleoptera, Lampyridae, Elateridae, DNA

Geographical variation in the nightly onset of flashing in the Japanese terrestrial firefly, *Luciola parvula*

TAKURO YASUOKA[1*], TSUYOSHI YAGI[1,2], KAORU MAETO[1] & MAKIO TAKEDA[1]

[1]*Graduate School of Agricultural Science, Kobe University, 1-1 Rokkodai, Nada-ku, Kobe, 657-8501, Japan.*
[2]*The Museum of Nature and Human Activities, Hyogo, 6 Yayoigaoka Sanda Hyogo 669-1546, Japan.*
**Email: tidatiki@yahoo.co.jp*

The firefly *Luciola parvula* is endemic to Japan. Our previous research revealed that *L. parvula* populations exhibited geographical variation in morphological and ethological characters such as body size and the time of night at which they flash. This variation may lead to reproductive isolation. Detailed documentation of this variation and the controlling mechanisms is needed to determine the taxonomic status of these populations and aid in their conservation. We conducted an overnight survey of the time of night at which flashing occurred at about 60 sites. The geographic groups mainly split into two based on their time of flashing: a dusk-flashing type (flashing actively over a period of about 2 hours soon after sunset), and a midnight-flashing type (flashing actively only after more than 4 hours after sunset). The dusk-flashing type is distributed in the Tohoku, western Honshu, Shikoku and Kyushu areas. The midnight-flashing type is distributed in the central Honshu and surrounding coastal areas. There were, however, some populations that flashed at intermediate times, and their geographical distributions lay between those of the dusk- and midnight-flashing types. There was no significant correlation between time of flashing and temperature, humidity or ambient light at the respective sites on the night of research. There was also no statistically significant effect of average yearly temperature, rainfall, solar radiation and depth of snow on time of flashing. Males that flash at different times continued to do so under laboratory conditions. These results suggest that the difference in time of flashing was genetically determined.

Keywords: Firefly, *Luciola parvula*, flash behaviour, geographical variation, Japan

Explanatory research using novel luciferase genes from Southeast Asian fireflies under the Convention of Biological Diversity

Ryutaro Akiyoshi*, Katsunori Ogoh & Hirobumi Suzuki

Olympus Corporation, 2-3 Kuboyama-cho, Hachioji-shi, Tokyo 192-8512, Japan.
**Email: ryutaro_akiyoshi@ot.olympus.co.jp*

The luciferin-luciferase bioluminescence system (L-L system) is widely used for cell function analysis in the life sciences. Cellar information mediated by L-L system is detected by a photosensor as light intensity. As light emitting from cells is measured in integrated value through all cells, it is impossible to analyse the cellar information from cells individually in images of tissue or embryos. To resolve this, we developed the luminescence microscope (LV200, Olympus) and used it for gene expression analysis in embryogenesis. However, commercially available luciferase expression vectors are inadequate for imaging analysis at the cell level because of weak luminescence. We therefore searched for a brighter luciferase gene in Southeast Asian fireflies, observing the Convention of Biological Diversity. We report an attempt to conduct explanatory research using novel firefly luciferase genes from Southeast Asia, and the application of bioluminescence imaging for gene expression analysis during embryogenesis in *Drosophila*. We cloned the armadillo promoter region from the *D. melanogaster* genome and constructed the plasmid vector armadillo::Luciferase. The armadillo is a homolog of vertebrate catenin β1 and is involved in many processes during embryogenesis, such as cell-cell adhesion, establishment and maintenance of cell polarity, and development of the embryonic nervous system. The armadillo::Luciferase vector was injected into a fertilised egg of *D. melanogaster*, and we succeeded in observing gene expression pattern of the armadillo during embryogenesis using the luminescence microscope. The bioluminescence imaging assay provides a new research method for observing not only gene expression pattern but also morphological changes during developmental processes.

Keywords: Embryogenesis, bioluminescence, luminescence microscope, luciferase, genes, Convention of Biological Diversity

The firefly population and other insects along the banks of the Johor River in Malaysia

S. Norela[1*], M.H. Shahril[2], A. Mohd Noh[2] & A. Maimon[1]

[1]*School of Environmental and Natural Resource Sciences, Faculty of Science and Technology, Universiti Kebangsaan Malaysia, 43600 Bangi, Selangor, Malaysia.*
[2]*TNB Research Sdn Bhd, No. 1, Jalan Air Hitam, Kawasan Penyelidikan, 43000 Kajang, Selangor, Malaysia.*
**Email: vozela@ukm.my*

The population of the firefly *Pteroptyx tener* and the diversity of insects along the banks of the Johor River in Kota Tinggi, Johor were examined by field sampling. Environmental parameters were also measured in the field, i.e., air temperature, wind speed, relative humidity and light intensity. The order Coleoptera was dominant in samples taken during both the new and full moon phases, but the firefly *P. ten*er was more abundant during the new moon than during the full moon. The fireflies were found to be abundant only on certain species of riparian trees, particularly *Sonneratia caseolaris* and *Gluta renghas*. The mixed vegetation along the riverbank provides good habitat for *P. tener* and other arthropods. Cantharidae, Cerambidae and Chrysomelidae were the most abundant familes of Coleoptera in all samples. Tetrigidae were common in the order Homoptera and Crambidae and Sesiidae in the order Lepidoptera. Bombylidae, Simuliidae and Megamerinidae were the families found from the order Diptera, while Formicidae predominated in the order Hymenoptera. These families were collected on all sampling occasions and constituted the predominant invertebrate biota associated with vegetation inhabited by the fireflies. Insect diversity was higher during the new moon than during the full moon (Shannon-Weiner diversity index H' = 1.56 and 1.29, respectively). However, there was no significant correlation between the environmental factors measured and the abundance of fireflies during either moon phase. Recommendations are also made on conservation of the Johor River ecosystem to help protect the riverside habitats in which these insects live.

Keywords: Firefly, *Pteroptyx tener*, insect diversity, moon phase, Johor River, Malaysia

Comparison of firefly larval abundance in three different landscapes along the Selangor River, Malaysia

B. Nada[1*], L.G. Kirton[1] & Y. Norma-Rashid[2]

[1]*Tropical Forest Biodiversity Centre, Forest Research Institute Malaysia, 52109 Kepong, Selangor, Malaysia.*
[2]*University of Malaya, 50603 Kuala Lumpur, Malaysia.*
**Email: nada@frim.gov.my*

Pteroptyx tener is the most abundant firefly along the Selangor River near the coastal town of Kuala Selangor in Malaysia. Various studies have been conducted on the adult stage of the firefly, especially since the insect has become a tourist attraction that contributes to the socio-economic welfare of the local community. However, in-depth research on the habitat requirements of the larval stage of the firefly has been greatly lacking, in spite of the fact that more than 60% of the life span of the firefly is spent in this stage. The larvae are also more vulnerable as they are terrestrial, inhabiting the floodplains of the river, making them more easily affected by the quality of river water and changes in the landscape due to human activity. This study focuses on quantifying larval abundance of *P. tener* in three major landscapes, that is, naturalised sago stands, cultivated oil palm and village orchards, along a stretch of the Selangor River inhabited by the firefly. The abundance of the firefly larvae and potential host snails was determined from five sites representing each landscape, which were sampled at night by active-searching in quadrats placed randomly along transects. It was found that the firefly larvae and host snail were most abundant in naturalised sago stands, while the two cultivated landscapes were found to be unsuitable as habitats in which the fireflies could breed. Physical and environmental characteristics of the three landscapes were assessed and related to the abundance of the firefly larvae and the snails.

Keywords: Firefly, larvae, *Pteroptyx tener*, habitat, Selangor River, Malaysia

Monitoring populations of aggregating fireflies using digital image analysis †

Laurence G. Kirton*, Veronica Khoo & B. Nada

Tropical Forest Biodiversity Centre, Forest Research Institute Malaysia, 52109 Kepong, Selangor, Malaysia.
**Email: laurence@frim.gov.my*

The use of digital night photography and image analysis for monitoring populations of fireflies is described based on its use in monitoring the population of *Pteroptyx tener* along the Selangor River in Malaysia. The technique enables the monitoring of large areas of habitat that would not be possible to monitor with conventional methods. A single lens reflex camera, with a good sensor, equipped with a telephoto zoom lens mounted on a sturdy tripod placed 60-270 m from trees across the riverbank, was used to capture images of flashes of the fireflies at night. The flashes appeared as faint bright spots against a dark background. Gamma levels of images were adjusted to enable visual interpretation, and bright spots were counted. Exposure time was set to capture just two flash cycles to reduce the likelihood of registering multiple bright spots from each firefly due to movement. Techniques for ensuring reproducible image composition at night and selecting appropriate camera settings are described, as well as the use of software-aided image analysis to estimate counts of bright spots. Standardisation of time of image capture and camera position is important. Short term temporal fluctuations in flashes need to be considered in the design of replication. Calibration is required to make results comparable across different operators or cameras, and the limitations of image analysis in differentiating the weak light signals of flashing fireflies from ambient light need to be considered. The technique can also be adapted to monitor populations of any light producing insects.

Keywords: Firefly, population, monitoring techniques, photography, image analysis, *Pteroptyx tener*

† Published as: Kirton, L.G., Nada, B., Khoo, V. & Phon, C.-K. 2012. Monitoring populations of bioluminescent organisms using digital night photography and image analysis: a case study of the fireflies of the Selangor River, Malaysia. *Insect Conservation and Diversity* **5** 244-250. First published online July 2011, doi: 10.1111/j.1752-4598.2011.00157.x (with 4 pages of online supporting information).

Lampyrid 2012. **2**:197-204

Poster Abstracts

Pteroptyx macdermotti McLean — the unusual *Pteroptyx* firefly

Lesley A. Ballantyne[1*], Tuan-Wah Wong[2] & Colleen A.L. Goh[3]

[1]*School of Agricultural and Wine Sciences, Charles Sturt University, PO Box 588, Wagga Wagga 2678, Australia.*
[2]*National Parks Board, Raffles Building, SBG, 1 Cluny Rd, Singapore 259569, Singapore.*
[3]*31 Hindhede Walk, #03-12, Singapore 587967, Singapore.*
**Email: lballantyne@csu.edu.au*

P. macdermotti was described in Ballantyne and McLean (1970) from 4 males and a possible female from three locations in the Philippine Islands (Leyte, Samar and Palawan). These are the only published records, and until now this was regarded as a rare species. It is unusual and of great interest in phylogenetic studies as it appears to exhibit features of both *Pteroptyx* and *Pyrophanes* while having much reduced light organs in the terminal abdominal segment, which has apical modifications of unusual structure, the use of which until now has not been explained. A new record from Donsol, southern Luzon indicates that not only might the species be quite prolific there but the opportunity may exist to investigate female and larval biology as well as male flashing patterns.

Keywords: Firefly, *Pteroptyx macdermotti*, distribution, Philippines, morphology, new record

Discovery of synchronous flashing in a new species of *Luciola* from Europe (Paxos, Ionian Islands, Greece)

Raphaël De Cock[1*], Andrew Moiseff[2], John Day[3] & Shelley Castle[4]

[1]*Evolutionary Ecology Group, University of Antwerp, Belgium.*
[2]*Department of Physiology & Neurobiology, University of Connecticut, USA.*
[3]*Centre for Ecology & Hydrology – Wallingford, Crowmarsh Gifford, Wallingford, Oxon, U.K.*
[4]*University College Falmouth incorporating Dartington College of Arts, Devon, U.K. in collaboration with the School of Biological Sciences, Bristol University, U.K.*
**Email: redecock@hotmail.com*

Synchronous flashing behaviour is demonstrated for the first time in a European *Luciola* species from the Greek Ionian Island of Paxos, a behaviour previously only known in some species of fireflies from North America, Southeast Asia and Oceania. The mitochondrial cytochrome oxidase I (COI) gene, the standardised locus for DNA barcoding and molecular taxonomy, was amplified and sequenced from *Luciola* specimens from Paxos and Sardinia and compared to a sequence dataset of *Luciola lusitanica* and *L. italica* specimens from Italy. Although the species morphologically resembles the well-known, non-synchronous flashing southern European species *L. lusitanica*, the genetic analysis strongly suggests that the *Luciola* from Paxos and Sardinia are a different species from *L. italica* and *L. lusitanica*. Analysis of video recordings demonstrates that the synchronous flashing is maintained over time and that the underlying rhythm is relatively constant. Therefore, both differences in flash behaviour as well as genetic analysis provide support that the *Luciola* from Paxos is a new species.

Keywords: Firefly, *Luciola*, new species, synchronous flashing, DNA, Paxos

The application of digital photography to study group display behaviour in aggregating fireflies — a preliminary test based on nocturnal activity rhythm of *Pteroptyx tener*

Veronica Khoo* & Laurence G. Kirton

Tropical Forest Biodiversity Centre, Forest Research Institute Malaysia, 52109 Kepong, Selangor, Malaysia.
**Email: veronica@frim.gov.my*

Flash behaviour of synchronous fireflies has been a subject of much study. Conventionally, devices such as photomultipliers and oscilloscopes as well as video recordings are used, which can be analysed to provide information on overall light intensity over time. In this study, we demonstrate the use of digital photography as a method of studying group flash behaviour that provides information on the numbers of fireflies flashing at any one time. This method involves the use of a SLR camera with a highly sensitive sensor and a zoom lens, and produces images with bright spots against a dark background, which are amenable to analysis using particle detection by image analysis software. The method is tested in relation to the nocturnal activity rhythms of the congregating, synchronously flashing firefly *Pteroptyx tener*. Two cameras were placed on a single tripod with the main camera equipped with a telephoto (zoom) lens focusing on an area dense with firefly flashes and taking images at 15 s intervals, while the second camera equipped with a wide-angle lens captured the whole stretch of display trees visible across the riverbank at 60 s intervals. A total of 2008 images were captured from 8.30 pm to 4.30 am on the main camera on a single night. The flashes registering on these images were analysed using image analysis software and averaged over 15 minute intervals. The average number of counts was plotted against the time of night. The graph showed that the number of fireflies flashing fluctuated somewhat throughout the night but was at its peak early in the night and began to decrease as the night progressed, becoming lower around 10.30 pm and very low in the early hours of the morning. Visual inspection of the images taken by the second camera corroborated the results of the first camera over a wider area of the riverbank vegetation. This study demonstrates that digital photography can be a valuable tool to supplement other methods of analysing group behaviour of fireflies, and it has the advantage of enabling differentiation at the level of individual fireflies unlike methods that measure group effects alone.

Keywords: Firefly, flash behaviour, photography, *Pteroptyx tener*, activity rhythm

Presence and distribution of Spanish glow-worms based on a survey by means of a photo-biodiversity database

J.R. Guzmán Álvarez[1]* & Raphaël De Cock[2]

[1]*Department of Forestry, University of Córdoba, Spain.*
[2]*Evolutionary Ecology Group, University of Antwerp, Belgium.*
**Email: gusanosdeluz@gmail.com*

Up till today, there is still a lack of knowledge of Iberian fireflies, except for what is known from recent studies of Portuguese Lampyridae and a description of a new species, *Lampyris iberica*. A general revision in the Catalogue of Palaearctic Coleoptera and other references show the presence of at least 11 species of Lampyridae in Spain and Portugal: *Lampyris noctiluca*, *L. raymondi*, *L. iberica*, *Nyctophila reichii*, *N. heydeni* (only on the Balearic islands), *Lamprohiza mulsanti*, *L. paulinoi*, *Phosphaenus hemipterus*, *Luciola lusitanica*, *Phosphaenopterus metzneri* (no recent records) and *Pelania mauretanica* (no recent records). In spite of the fact that glow-worms are relatively common in Spain, especially in the countryside where villagers are used to the presence of these insects, studies of the Spanish firefly fauna are very lacking and mainly limited to a few anecdotal records. However, as pointed out in other studies, there is a general perception that the number of glow-worms has decreased over the years. In order to improve knowledge of the lampyrid fauna of Spain, the project: "Have you seen a glow-worm?" was launched on the Web in May 2009 (http://www.gusanosdeluz.es). Soon after the launch of the site, the authors were invited to collaborate in a photo-biodiversity database called "Biodiversidad Virtual," acting as authorities on lampyrids in identifying the species uploaded by website members. From May 2009 to May 2010, 21 online forms were received and 159 photos were examined (106 of *Nyctophila reichii*, 44 of *Lampyris* spp. and 9 of *Lamprohiza* spp.). Identification of larvae is difficult as there are no clear characteristics that discriminate between juveniles of the three species of *Lampyris* recorded in Spain. Four *Lampyris iberica* adults have been recorded and at least one adult male of *Lampyris noctiluca* from the North of Spain. Interestingly, *Nyctophila reichii* shows a much more Mediterranean and southern distribution than the *Lampyris* species. *Lamprohiza* shows a very distinct presence in the east Mediterranean area along the Valencian, Catalan and Murcian coastlines, with the exception of a record from the north-west in Galicia, in the A Coruña province. Although it may still be too early to be sure, a possible explanation for this pattern may be the existence of two different species of *Lamprohiza*. The characteristics of the eastern *Lamprohiza*, in particular, suggest that it is a new species.

Keywords: Glow-worm, survey, photo database, Spain

Distribution and habitat of *Asymmetricata circumdata* (Motsch.) (Coleoptera: Lampyridae: Luciolinae) in the north of Thailand

SORASAK NAK-EIAM* & WANDEE WATTANACHAIYINGCHAROEN

Department of Biology, Faculty of Science, Naresuan University, Phitsanulok 65000, Thailand.
**Email: aimbryo@yahoo.com*

An investigation of the distribution and habitat of *Asymmetricata circumdata* (Motschulsky) in the north of Thailand was conducted from July 2007 to August 2009. Fifteen study sites comprising different geographical areas and habitats were surveyed. The altitude of the study areas ranged between 38 and 1,072 m a.s.l. Collection of adult *A. circumdata* was carried out between 6.00 pm and 12.00 am using insect sweep nets. Results showed that *A. circumdata* had a broad distribution in the north of Thailand, and was found throughout summer and the rainy season. It was most abundant in mixed deciduous forests but also found in disturbed habitats such as urban and cultivated areas. However, it was more abundant in natural habitats, with more than 50 individuals collected per study site. In agricultural and urban areas, less than 5 individuals were obtained per study site. This shows that even though *A. circumdata* can occur in a range of habitats, they prefer natural habitats. Natural habitats therefore have to be protected for the survival of *A. circumdata*. A distribution map of *A. circumdata* in the north of Thailand is provided.

Keywords: Firefly, *Asymmetricata circumdata*, distribution, habitat, Thailand

The synchronous fireflies of Petchaburi River estuary

SURACHAI CHOLDUMRONGKUL[1*], NANTIYA RATANACHAN[2] & NONGPANGA PACHEY[2]

[1]*Royal Forest Department, 61 Phaholyathin Road, Ladyao, Jatujak, Bangkok 10900, Thailand.*
[2]*Forest Entomology and Microbiology Section, Forest Research and Development Office, Department of National Parks, Wildlife and Plant Conservation, Bangkok, Thailand.*
**Email: surach94@yahoo.com*

A survey of the synchronous fireflies of Petchaburi River was conducted at Pak Klong canal, Baan Laem District, Petchaburi province from August 2008 to 2009. Fireflies were surveyed once a month during the new moon phase of the month, from 8.00–9.00 pm. At least 3 display trees were sampled, firefly specimen collected, the number of individuals counted and all display tree species recorded. Species richness, abundance and density were analysed. There were two species of *Pteroptyx* co-existing on the same display trees, that is, *P. malaccae* and *P. valida,* which occurred in a ratio of 9:1 respectively. The greatest abundance was in March 2009 while the density was highest in August 2008. The species congregated mainly on *Sonneratia caseolaris* trees along the river bank.

Keywords: Firefly, Petchaburi River estuary, Thailand, *Pteroptyx malaccae*, *Pteroptyx tener*

Discovery of an additional species of firefly along the Selangor River, Malaysia based on morphologically distinct larvae

B. NADA[1*], LESLEY A. BALLANTYNE[2], LAURENCE G. KIRTON[1] & Y. NORMA-RASHID[3]

[1]*Tropical Forest Biodiversity Centre, Forest Research Institute Malaysia, 52109 Kepong, Selangor, Malaysia.*
[2]*Charles Sturt University, P.O. Box 588, Wagga Wagga, Australia.*

[3]*University of Malaya, 50603 Kuala Lumpur, Malaysia.*
**Email: nada@frim.gov.my*

Previous studies have documented the occurrence of three firefly species along the Selangor River in Malaysia. However, these records were based on adult fireflies, and no attempt has been made to identify the different larvae occurring in the area. During sampling conducted for a firefly larval habitat study along this river, three morphospecies of firefly larvae were found. Systematic sampling was conducted in sago stands, oil palms plantations and village orchards. Larvae collected comprised *Pteroptyx tener*, which was by far the most numerous species, and two other species of uncertain identity, referred to here as sp. 1 and sp. 2, which occurred at low densities. The morphology and possible identities of these two species are discussed together with the morphology of the commoner *P. tener*. Based on the presence of a single male *Pteroptyx valida* at a site where sp. 1 was collected, it is possible that this species is *P. valida*, which can also be seen in very small displays of adults along the Selangor River. The remaining species, sp. 2, which was found mainly in orchards, does not resemble *Pteroptyx malaccae*, for which adults have been reported from the area, although without substantiating specimens. The only specimen successfully bred to adult stage was a female that cannot be reliably identified without a male. It however, clearly indicates the discovery of an additional species along the Selangor River, the exact identity of which remains to be discovered.

Keywords: Firefly, larva, morphology, Selangor River, Malaysia

Protecting the natural attractions of a historic riverine village in Negeri Sembilan, Malaysia

N.R. Hashim*, W.F.A.W. Jusoh, A.M. Mohd Shahrudin, N. Mohd Fazlin, M.N. Mohd Nassrul Safre, M. Rathiveerachelvi & S. Alik

Heritage Colony, Universiti Putra Malaysia, 43400 Serdang Selangor.
**Email: rasidah@env.upm.edu.my*

The conservation of highly threatened habitats such as mangrove forests is difficult because of development pressures. But when the local people are interested and fully engaged in a conservation initiative, the future of such habitats may not be so bleak. In the historic riverine village of Kampung Sungei Timun in Negeri Sembilan, Malaysia, the villagers realise that the river and the surrounding mangrove forests need to be sustainably utilised and developed for their economic and heritage values as well as for their biological diversity. Some villagers are directly involved in ecotourism by offering river cruises. Others set up shops selling home cooked food to visitors. Besides the villagers, other stakeholders (the state government, visitors and researchers) can also play active roles in the conservation of this historic mangrove village. The Kampung Sungei Timun area is rich in Malay history. According to local folklore, several centuries back a Sumatran Prince, Raja Melewar, moored his sampan at the village's jetty while on his way to Seri Menanti to become the Ruler of Negeri Sembilan. Although Kampung Sungei Timun and its surrounding areas have been exposed to long-term anthropogenic disturbances because the history of human settlement started much earlier than the 17th century, some interesting wildlife species such as fireflies and crocodiles that are sensitive to habitat degradation can still be found there. However, judging from the types and extent of land use around the village, such as oil palm plantations and poultry industry, the mangrove habitats are already threatened by development. And although wildlife hunting is relatively unknown in Kampung Sungei Timun, the wildlife population could be wiped out by further destruction of the mangrove habitat as well as by river pollution. When this happens, the mangrove forest reserve located downstream of Kampung Sungei Timun will be the last remaining wildlife habitat in the area. Since December 2007, our research group (Heritage Colony) has been conducting several related studies in our effort to provide much needed knowledge on the flora and fauna of the mangrove ecosystem. The findings and implications of our research will be continuously made available to the villagers and the state government in order to inform these stakeholders of the conservation importance of the site.

Keywords: Mangrove, conservation, wildlife, Negeri Sembilan, Malaysia

Enhancing the firefly population at Pasir Ris Park, Singapore through active management of mangroves

Y. Mohamad

Blk 497J 05-74, Tampines Street 45, Singapore 527497.
Email: Mohamad_YUSOFF@nparks.gov.sg

The creation of Pasir Ris Park in Singapore preserved a 6 ha mangrove, a remnant of its once sprawling coverage. In 2000, a group of researchers found less than a dozen fireflies near Bakau Trail, deep inside the mangrove – a disappointing find. Several surveys identified factors inhibiting fireflies from breeding: pesticide application, habitat loss and lack of tidal inundation. However, little was done at the time to address these problems. In 2008, a group of National Parks Board officers spotted fireflies at the Bakau and Api Api trails, reigniting interest in protecting its habitat and enabling the insect to breed in situ. Further research revealed how to create conducive natural breeding grounds. Park lights may affect firefly mating behaviour, so plants were grown to act as a screen that would block light from penetrating the mangroves. Fogging activities indiscriminately killed the fireflies and crippled the food supply to their larvae (*Littoraria* snails) – these were stopped. Night surveys also showed that larvae clustered on the wet surface around *Nypa fruticans*, thus we left leaf litter uncleared to gradually decompose and attract snails. Park workers carried out trenching along the riverbank of Sungei Tampines to reinvigorate mangrove growth via sea water rivulets. This was of utmost importance to prevent freshwater-loving plants from invading the ecosystem and crowding out true mangrove plants. Freshwater retards mangrove proliferation. Subsequently, thirty *Sonneratia caseolaris* trees were reintroduced, adding to the mangrove's species diversity and providing additional display trees for adult *Pteroptyx valida*. A recent survey conducted in January 2010 revealed a healthy population of fireflies. They can now be observed on most nights just by walking on the boardwalk, indicating successful completion of the first phase of the repopulation project.

Keywords: Firefly, conservation, park management, Singapore

Firefly photography

Vor Yiu

Hong Kong Entomological Society, 31E Tim Sam Tsuen, Kam Sheung Road, Yuen Long, N.T., Hong Kong.
Email: yiuvor@hkentsoc.org

The light displays of fireflies in the dark are always spectacular. Using a camera to capture the phenomenon for publication can be an effective method of drawing the attention of the public to these creatures and highlighting the need to conserve them. When used properly, photography can also be a fast and effective way of recording morphological, biological and ecological information of fireflies for scientific studies. However, cameras are far less sensitive than the human eye and are not simple to use to achieve good results with fireflies. Specialised equipment and technique are necessary to make photographing fireflies possible. The form and intensity of light emissions by different firefly species and different stages of their life cycle are diverse. Different techniques are needed to photograph different fireflies. Results of different camera settings in different situations are demonstrated in this poster.

Keywords: Firefly, photography

Printed in Dunstable, United Kingdom

91342624R00123